Advances in Dielectrics

Series Editor

Friedrich Kremer, Fakultät für Physik und Geowissenschaften, Peter-Debye-Institut für Physik der weichen Materie, Universität Leipzig, Leipzig, Germany

Aims and Scope

Broadband Dielectric Spectroscopy (BDS) has developed tremendously in the last decade. For dielectric measurements it is now state of the art to cover typically 8-10 decades in frequency and to carry out the experiments in a wide temperature and pressure range. In this way a wealth of fundamental studies in molecular physics became possible, e.g. the scaling of relaxation processes, the interplay between rotational and translational diffusion, charge transport in disordered systems, and molecular dynamics in the geometrical confinement of different dimensionality – to name but a few. BDS has also proven to be an indispensable tool in modern material science; it plays e.g. an essential role in the characterization of Liquid Crystals or Ionic Liquids and the design of low-loss dielectric materials.

It is the aim of "Advances in Dielectrics" to reflect this rapid progress with a series of monographs devoted to specialized topics.

Target Group

Solid state physicists, molecular physicists, material scientists, ferroelectric scientists, soft matter scientists, polymer scientists, electronic and electrical engineers.

Jacopo Parravicini
Editor

Fröhlich Entropy Estimation and Dielectric Calorimetry

A Flexible Investigation Technique for Condensed Matter

Springer

Editor
Jacopo Parravicini
Dipartimento di Fisica & Astronomia
Università di Firenze
Sesto Fiorentino, Firenze, Italy

ISSN 2190-930X ISSN 2190-9318 (electronic)
Advances in Dielectrics
ISBN 978-3-032-21453-9 ISBN 978-3-032-21454-6 (eBook)
https://doi.org/10.1007/978-3-032-21454-6

This Springer imprint is published by the registered company Springer Nature Switzerland AG
The registered company address is: Gewerbestrasse 11, 6330 Cham, Switzerland

Foreword

Herbert Fröhlich: Life and Work

Herbert Fröhlich was born on December 9, 1905 in Rexingen, a small town in the Black Forrest; he entered the University of Munich as an undergraduate in 1927. Guided by Arnold Sommerfeld, he obtained the doctoral degree "D. Phil" (for a thesis on the photoelectric effect in metals) after only 3 years without ever having taken an undergraduate degree! From his first post as private lecturer at the University of Freiburg, where he was responsible for introducing modern physics, he was dismissed because of the upcoming Nazi regime and left Germany in 1933 for Russia, to work (at Frenkel's invitation), as a "Foreign Expert," in Joffe's Physico-Technical Institute in Leningrad (St. Petersburg), Russia. There, he became acquainted with current work on semiconductors, and included a discussion of them in his now famous book *Elektronentheorie der Metalle* (Springer, 1936), which for some years was the only textbook to contain a treatment of semiconductors. Later it proved very influential when the technological potential of these materials started to be appreciated, particularly in the USA, where it was reprinted (in the German original) in 1943. The political situation in Russia forced him to flee again, and in 1935, he joined Mott's department at the University of Bristol, England. There, he remained until 1948, having advanced to the position of a Reader. Recommended by J. Chadwick, he took up the first Chair of Theoretical Physics at the University of Liverpool—a position he held until his retirement in 1973, after which he was Professor Emeritus from 1976 until his death on January 23, 1991, at the age of 85. Between 1973 and 1976, he was Professor of Solid State Electronics at the University of Salford, though he still maintained an office in Liverpool, so he spent a total of 43 years there.

Fröhlich was elected Fellow of the Royal Society in 1951 and awarded with the Max Planck Medal of the German Physical Society in 1972; he received numerous honorary degrees worldwide. From 1974 until his death, he was an external member of the Max Planck Institute for Solid State Physics, Stuttgart, where he regularly

Fig. 1 Herbert Fröhlich photographed at a conference in Prague (Czech Republic) in the mid-1980s. (Courtesy of Prof. F. Kremer)

spent extended visits, as he did also to many other parts of the world, lecturing and discussing physics.

Fröhlich made fundamental contributions to various fields of physics, as diverse as meson theory, the theory of superconductivity, and the theory of dielectrics. He was the first who suggested that electron-phonon interaction is the key to explain superconductivity and his concepts paved the way to the approach of Bardeen, Cooper, and Shrieffer introducing "Cooper pairs." His book *Theory of Dielectrics* was for a long time—besides Debye's *Polar Molecules*—the standard textbook book discussing the interaction of electric fields with matter. In his later years, he became interested in order and dynamics in biological systems and coherent excitations therein. This triggered a multitude of experimental activities which are summarized in two textbooks *Coherent Excitations in Biological systems* (Eds. H. Fröhlich and F. Kremer, Springer 1983) and *Biological Coherence and Response to External Stimuli* (Ed. H. Fröhlich, Springer 1988).

This novel volume of Advances in Dielectrics aims to describe H. Fröhlich's concepts on thermodynamics and dielectrics in detail and to present experiments where it was fruitfully applied.

Leipzig, Germany
December 2025

Friedrich Kremer

Preface

This volume, entitled *Fröhlich Entropy Estimation and Dielectric Calorimetry* within the series Advances in Dielectrics, seeks to elucidate the features and implications of assessing variations in thermodynamic potentials as framed by classical Fröhlich thermodynamic theory. This experimental methodology, which so far has been developed predominantly through Fröhlich entropy estimation, constitutes a relatively simple and cheap procedure, which may be regarded as a viable alternative to more established and widely employed techniques.

Fröhlich's theory addresses the variations in thermodynamic parameters arising in an ideal dielectric material subjected to an external electric field. When this process is experimentally implemented and quantitatively assessed, this theory—under appropriate assumptions—enables the determination of thermodynamic quantities through dielectric quantities measured as functions of temperature. This framework thereby opens the way to various experimental advances, allowing a transition from conventional dielectric thermal analysis to the more specialized domain of *dielectric calorimetry*. In fact, among the thermodynamic variables altered by the presence of an electric field, the most significant is the variation of entropy, referred to as *Fröhlich entropy* from the name of H. Fröhlich, who first derived the relevant equations. When the imaginary component of the static dielectric function is negligible—namely, far from resonance conditions—Fröhlich entropy can be evaluated from the real part of the dielectric function. As emphasized by Fröhlich himself, this quantity may serve as a figure-of-merit for the degree of order within the physical system under consideration. Accordingly, when interpreted in this sense, the values of Fröhlich entropy obtained from experimental data provide valuable insights into the evolution of the investigated compound, particularly with respect to temperature dependence, phase behavior, stability, transitions, and order-disorder states and phenomena. Since the underlying assumptions of this approach are relatively weak, it can be applied to a wide range of physical systems, thereby serving as a valuable extension of standard dielectric measurement techniques, such as dielectric thermal analysis and spectroscopy. Indeed, *Fröhlich entropy estimation* has thus far been successfully employed across diverse areas of condensed matter physics, including dipolar and nematogenic liquids, dipolar glasses, organic molecular crystals and semiconductors, metallic nanoparticles, inorganic disordered ferroelectrics, as well as proteins and enzymes. This breadth of applications demonstrates that Fröhlich entropy estimation constitutes a reliable investigative tool, endowed with considerable potential that remains to be further exploited.

The structure of this volume is organized as follows. After a general introduction to the topic (chapter "Introduction: Dielectric Calorimetry and Fröhlich Entropy Estimation"), the fundamentals of classical dielectric thermodynamic theory and related applications are briefly presented, with particular emphasis on Fröhlich's contribution (chapter "Elements of Thermodynamics of Dielectrics and Basic Theory of Fröhlich Entropy"); the specific extension of this framework to anisotropic and nonlinear media is then addressed in chapter "Generalized Fröhlich Entropy: Anisotropic and Nonlinear Media". Subsequently, widely employed experimental techniques for analyzing the state of order are reviewed in chapter "Experimental Techniques to Investigate the State-of-Order in Condensed Matter", for the purpose of comparison. Chapter "Meaning and Applications of Fröhlich Entropy" examines the physical meaning of the Fröhlich entropy quantity, including a discussion of Fröhlich's interpretation and its relevance for experimental investigations. Finally, the practical application of Fröhlich entropy and theory to specific physical systems and conditions is explored, namely, metallic nanoparticles (chapter "Fröhlich Entropy of Nanoparticles: Investigation of Phases, Phase Transitions and Melting"), soft matter and extreme conditions (chapter "Fröhlich Entropy in Soft Matter and Extreme Conditions"), organic semiconductors (chapter "Dielectric Thermodynamics of Organic Semiconductors"), and ferroelectric perovskites (chapter "Order–Disorder Transitions of Ferroelectric Perovskites by Fröhlich Entropy Investigations").

The contributors to this collaborative book were selected from among researchers who have successfully applied this approach to investigations in condensed matter physics. It is my hope that the present work will encourage its broader adoption within the scientific community, thereby providing researchers with a versatile, rigorous, and powerful instrument of inquiry.

Sesto Fiorentino, Firenze, Italy
December 2025

Jacopo Parravicini

Acknowledgements My thanks go to all scientists who contributed to this volume, that are recalled in the List of Contributors. I particularly acknowledge GB. Parravicini, L. Lo Presti, R. Soave, M. Scavini, E. DelRe, P. Ghigna, P. Ben Ishai, Y. Feldman, and A. Sassella. Special thanks go to the Prof. F. Kremer for his great availability and his help and suggestions in the preparation of this book. I also thank E. Tombari and M. Mauri for useful discussions and suggestions.

Jacopo Parravicini

L. D.

Contents

Acronyms and Variables

ADPs	Atomic displacement parameters
BB	Bouchard and Brioli
BDS	Broadband dielectric spectroscopy
bR	Bacteriorhodopsin
$\mathbf{c}$	Electrostriction coefficient
c_D	Heat of polarization
C_p	Heat capacity at constant pressure
C_v	Heat capacity at constant volume
$\mathbf{d}$	Direct piezoelectric effect coefficient
$\mathbf{d}$	Converse piezoelectric effect coefficient
$\mathbf{D}$, D	Dielectric displacement field
DETA	Dielectric thermal analysis
DS	Diffuse scattering
DSC	Differential scanning calorimetry
DSE	Debye scattering equation
DTC	Differential thermal analysis
$\mathbf{E}$, E	Electric field
EBSD	Electron backscatter diffraction
EDS	Dispersive X-ray spectroscopy
ER	Ergodic relaxor (phase)
$\mathbf{f}$	Heat of deformation coefficient
F	Helmholtz free energy
F_E	Field-dependent Helmholtz free energy
F_0	Field-independent Helmholtz free energy
FD	Ferroelectric domain
FE	Fröhlich entropy
FR	Ferroelectric (phase)
FT	Fourier transform
G	Gibbs free energy
H	Enthalpy
HOMO	Highest occupied molecular orbital

I	Isotropic (phase)
K_B	Boltzmann constant
LUMO	Lowest unoccupied molecular orbital
N	Nematic (phase)
NEXAFS	Near-edge X-ray absorption fine structure
NR	Normal relaxor (phase)
NRM	Nuclear magnetic resonance
OMS	Organic molecular semiconductors
p	Pressure
$\mathbf{P}$, P	Density of polarization, polarization
P	Power
PAS	Positron annihilation spectroscopy
PDF	Pair distribution function
PE	Paraelectric phase
PND	Polar nanodomain
PNR	Polar nanoregions
q	Electrostriction coefficient
Q	Heat
RT	Room temperature
S	Entropy
s	Fröhlich entropy for unitary field
S_E	Fröhlich entropy
S_0	Field-independent entropy
SEM	Scanning electron microscopy
STM	Scanning tunneling microscopy
T	Temperature
T_B	Burns temperature
T_C	Curie-Weiss temperature
T_g	Glass transition temperature
TDS	Thermal diffuse scattering
TEM	Transmission electron microscopy
TEN	Tetracene
TG	Thermogravimetry
TGA	Thermogravimetric analysis
TL	Thermoluminescence
TM	Thermomagnetometry
TMA	Thermomechanical analysis
U	Internal energy
u_e	Density of electric energy
U_e	Electromagnetic energy
U_E	Field-dependent internal energy
U_0	Field-independent internal energy
v	Volume
WH	Williamson and Hall
XANES	X-ray absorption near edge structure

XAS	X-ray absorption
XRD	X-ray diffraction
4T	α-quaterthiophene
α	Thermal expansion coefficient
$\boldsymbol{\alpha}$	Thermal expansion coefficient
β	Thermomechanical coefficient
γ_E	Volumetric electrostriction
γ_p	Pressure coefficient of polarization
$\boldsymbol{\varepsilon}$, ε	Dielectric constant
ε_s	Static dielectric constant
ε_0	Vacuum dielectric constant
$\boldsymbol{\eta}$	Strain tensor
κ_T	Isothermal compressibility
$\mathbf{\Pi}$	Pyroelectric coefficient
$\boldsymbol{\sigma}$	Stress tensor
$\mathbf{\Sigma}$	Elasticity

Contributors

Stefano Americo CNR-SCITEC, Milano, Italy

Nanochemistry, Istituto Italiano di Tecnologia, Genova, Italy

Paul Ben Ishai Department of Physics, Ariel University, Ariel, Israel

Mauro Coduri Department of Chemistry, Università degli Studi di Pavia, Pavia, Italy

Eugenio DelRe Dipartimento di Fisica, "Sapienza" Università di Roma, Roma, Italy

Yuri Feldman Institute of Applied Physics, Hebrew University of Jerusalem, Jerusalem, Israel

Paolo Ghigna Dipartimento di Chimica, Università degli Studi di Pavia, Pavia, Italy

Friedrich Kremer Peter Debye Institute for Soft Matter Physics, Faculty of Physics and Earth System Sciences, Universität Leipzig, Leipzig, Germany

Leonardo LoPresti Department of Chemistry, Università degli Studi di Milano, Milano, Italy

Giovanni Macetti Department of Chemistry, Università degli Studi di Milano, Milano, Italy

GianBattista Parravicini Dipartimento di Fisica, Università degli Studi di Pavia, Pavia, Italy

Jacopo Parravicini Dipartimento di Fisica & Astronomia, Università degli Studi di Firenze, Sesto F., Firenze, Italy

Yuwen Qiao School of Physics, Harbin Institute of Technology, Harbin, China

Adele Sassella Dipartimento di Scienza dei Materiali, Università degli Studi di Milano-Bicocca, Milano, Italy

Marco Scavini Department of Chemistry, Università degli Studi di Milano, Milano, Italy

Luca Sironi Department of Chemistry, Università degli Studi di Milano, Milano, Italy

Department of Chemistry, Durham University, Durham, UK

Raffaella Soave CNR-SCITEC, Milano, Italy

Peng Tan School of Physics, Harbin Institute of Technology, Harbin, China

Hao Tian School of Physics, Harbin Institute of Technology, Harbin, China

Martin Tress Molecular Photonics group, Peter Debye Institute for Soft Matter Physics, Faculty of Physics and Earth System Sciences, Universität Leipzig, Leipzig, Germany

Margherita Vacchini Department of Chemistry, Università degli Studi di Milano, Milano, Italy

Zijian Zhang School of Physics, Harbin Institute of Technology, Harbin, China

Introduction: Dielectric Calorimetry and Fröhlich Entropy Estimation

Jacopo Parravicini

Abstract The basic concepts on which this book is based are provided. After some general notes on the theory of thermodynamic for dielectric materials, a brief landscape about thermal analysis and calorimetric techniques is provided. Hence, the fundamental elements of the Fröhlich's thermodynamics theory for dielectrics are outlined in broad terms, arriving to the Fröhlich's thermodynamic equations for dielectric materials. Finally, the basic elements of Fröhlich entropy estimation techniques are described.

1 Overview

In most physics, chemistry, and material science books, thermal and electric (and magnetic) properties of materials are treated in isolation from the others. Even more so, this is true in treatises dealing with investigation techniques: the usual approach is to present thermal-based and electromagnetic-based techniques in well-separated discussions. Conversely, the real physical properties are strictly interrelated. hence a realistic treatise should also take into account the interrelations among physical quantities of such different nature. This should be all the more carried out nowdays, when the present physical sciences are showing increasing interests for developing cross-linked concepts, approaches, and paradigms. Usually, ideas start from some common ground among branches, to generate crossed ramifications. In physics, one of the most general concept, perhaps the most general at all, underlying all branches and topics, is the energy. This is also highlighted by the fact that, when a theory is systematically exposed, energetic concepts are often the starting points from which the discussion is developed. In condensed matter physics, we may recognize an apparently trivial phenomenon: when a sample of material is subjected to an electric field, its energy content changes. This implies that all its specific thermodynamics quantities display a variation. This is a basic assumption pointing out that electric

J. Parravicini (✉)
Dipartimento di Fisica & Astronomia, Università degli Studi di Firenze, Sesto F., Firenze, Italy
e-mail: jacopo.parravicini@unifi.it

J. Parravicini (ed.), *Fröhlich Entropy Estimation and Dielectric Calorimetry*, Advances in Dielectrics, https://doi.org/10.1007/978-3-032-21454-6_1

and thermal physical quantities are somehow correlated. Actually, roughly 100 years ago observations pinpointed that applying electric field to a crystal can produce a variation of its temperature. This is known as *electrocaloric effect*, which has been the first physical phenomenon where a specific thermodynamics of dielectric has been discussed. Actually, the studies on this effect dates back to the end of 1920s: between the World Wars and in 1950s the basic equations of this effect were provided [1]. In fact, this effect first highlighted the cross coupling between thermal and electrical physical variables. Without taking into account mechanical quantities, the basic relationships can be obtained starting from the Gibbs free energy $G = G(T, E)$ [2], i.e.[1]

$$dG = -SdT - DdE; \tag{1}$$

so we obtain the corresponding Maxwell's thermodynamic equations for a dielectric

$$\begin{aligned} S &= -\left(\frac{\partial G}{\partial T}\right)_E \\ D &= -\left(\frac{\partial G}{\partial E}\right)_T. \end{aligned} \tag{2}$$

If we consider a materials which displays a dielectric displacement D (or a polarization P), from (2) we can obtain the induced ΔS variation of entropy [1, 3, 4]

$$\Delta S = -\int_{E_1}^{E_2} \left(\frac{\partial D}{\partial T}\right)_E dE \tag{3}$$

so that the temperature variation is

$$\Delta T \propto -\int_{E_1}^{E_2} T \left(\frac{\partial D}{\partial T}\right)_E dE. \tag{4}$$

Equations (3) and (4) are among the oldest specific mathematical relationships where dielectric and thermodynamic quantities are mixed [1]. In fact, the first discussion about the specific thermodynamic behavior of dielectric materials was carried out in the context of such electrocaloric effect. Later on, such discussion largely grew up dealing several other phenomena such as ferroelectricity, pyeroelectricity, thermoelectricity, piezoelectricity [3–5]. When dielectric materials are specifically considered, there are a number of seminal works which develops the general theory of dielectrics [6–12]: often they start from basic energetic considerations, that is, in the macroscopic realm, from the basic thermodynamics of

[1] Where, as usual, S is the entropy of the material, T is the temperature. E is the applied electric field and D is the dielectric displacement vector.

a dielectrics. The first exhaustive theory, although considering quite idealized cases, was provided by H. Fröhlich, that specifies the basic equations [6, 7]. Following Fröhlich, we consider an ideal homogeneous isotropic dielectric, whose the total energy, the entropy, and the Helmholtz free energy, are U, S, F respectively. If the dielectric is subjected to a static electric field, in usual conditions, without loosing generality, we may write

$$U(T, E) = U_0(T) + U_E(T, E), \tag{5a}$$

$$S(T, E) = S_0(T) + S_E(T, E), \tag{5b}$$

$$F(T, E) = F_0(T) + F_E(T, E), \tag{5c}$$

where U_0, S_0, F_0 are respectively the variation of total energy, entropy, Helmholtz free energy at zero field, while U_E, S_E, F_E are their respective field-dependent parts.[2] The crucial point is to provide an analytical relationships of (5), making explicit the functional dependence of the involved thermodynamic potentials (dependent variables) on the temperature and dielectric function (independent variables). To accomplish such task, the development of a theory and a model, with suitable assumptions, is a mandatory ingredient. This is what is done by Fröhlich for the first time, whose procedure we will extensively expose in Sect. 3. The explicit formulation of (5) leads to some consequences. First, it allows a formal description of the correlations between thermodynamic evolution and dielectric features of a material. This can be useful, but, from an experimental perspective, this rigorous mathematical treatment has larger implications. Actually, it opens the possibility of building up a *dielectric calorimetry*, i.e. to obtain calorimetric information through dielectric measurements. This idea can be deduced by the interpretation provided by Fröhlich himself in his treatise, that we will report later.

2 Calorimetry, Thermal Methods, and Dielectric Thermal Analysis

2.1 Experimental Techniques for Thermal Investigation

There are a huge amount of techniques to investigate materials. Many of them are usually shared by physics, chemistry, material science and engineering, geology, biology. Each experimental technique and approach exploits specific physical principles, so it can be employed on suitable compounds or classes of samples and in specific conditions; moreover, it is able to highlight peculiar phenomena, which can, or can't, be visible depending on the used practical approach. Usually,

[2] We here neglect the vectorial features of the electric field. We are discussing such aspects in the next chapter "Generalized Fröhlich Entropy: Anisotropic and Nonlinear Media".

in natural sciences such as physics or chemistry, whenever experimental method requires that the sample to be investigated is perturbed. This is done by externally varying (at least) one of the physical quantities which are supposed to influence it, and, consequently, by measuring the correlated response of the sample. Often we may employ as independent variable the energy: this is done, e.g., in usual spectroscopy techniques such as photoluminescence, spectrophotometry, broadband dielectric spectroscopy, optical spectroscopy, Raman spectroscopy, etc. [13, 14]. Strictly correlated to this approach, thermal quantities, namely the temperature, can be assumed as driving variables. If we consider the large set of physical variables affecting a material sample, thermal quantities are usually among the simplest to be considered. In most cases heating (or cooling) is among the easiest procedure to carry out whenever it has to be examined. A lot of information about the nature of the material can be obtained by observing the corresponding behavior and quantitatively measuring the changes as it is heated (or cooled). In the investigation approach of the condensed matter physics, temperature and heat of the sample are externally modified in conditions without occurring any chemical reaction.

The scientific study of how heating alters sample characteristics led to the development of thermal analysis techniques [15, 16]. Mainly for historical reasons, *calorimetry* and *thermal methods* are usually considered as separate: the first one is the experimental technique measuring the amounts of heat, while the second ones are various methods giving the temperatures of specific thermodynamic phenomena, such as phase changes, decomposition, products of chemical reactions, etc. Although a classical definition of *calorimetry* would state that it is *the measurement of the heat changes occurring during some thermodynamic process* [15], we adopt such definition from a broader perspective. In fact, from the point of view of physicists, the description of a macroscopic system can be done through any suitable function belonging to the set of thermodynamic potentials, among which heat and total energy are included [17, 18]. Accordingly, by *calorimetry* we here mean a thermal analysis that results in the measurement of changes *in any macroscopic thermodynamic potential* describing the physical system under investigation. In this perspective, calorimetric studies may provide thermodynamic quantities that are expressed in energy-like units of measurement, such as heat, internal energy, free energy, entropy, enthalpy, etc. (see e.g. Table 1).

There are several techniques falling within that definition: they exploit the variation of thermal quantities, whose temperature is used in most cases as the independent variable to be spanned, in order to detect the corresponding variations of material-dependent physical quantities, that are meant as dependent variables. A brief summary of such experimental techniques is provided in Table 2. However, calorimetric methods usually measure changes in the physical properties of the studied specimen as a function of thermodynamic changes, especially temperature and heat. Among these properties we may find very heterogeneous features, such as colour, shape, size, mass, density and mechanical strength. There are also properties depending on molecular structure and nature of the material, such as heat capacity, enthalpy and entropy and also the structural and molecular properties which can be investigated by temperature-resolved X-ray and neutron diffraction, NMR, and

Table 1 Usual symbols and units of physical quantities most employed in calorimetric and thermal techniques

Physical quantity	Usual symbol	Unit	Abbreviation
Temperature	T	Kelvin	K
Pressure	p	Pascal	$\mathrm{Pa} = \mathrm{Nm}^{-2}$
Volume	v	Cube meters	m^3
Power	$\mathcal{P}$	Watt	$\mathrm{W} = \mathrm{Js}^{-1}$
Heat	Q	Joule	J
Heat capacity			
at constant pressure	C_p	Joule/kelvin	JK^{-1}
at constant volume	C_v	Joule/kelvin	JK^{-1}
Internal energy*	U	Joule	*J*
Helmholtz free energy	$F^{\dagger}$	Joule	J
Gibbs free energy	$G^{\ddagger}$	Joule	J
Enthalpy	H	Joule	J
Entropy	S	Joule/kelvin	JK^{-1}

* Also called *total energy*
† Also indicated with A
‡ Also indicated with Φ

Table 2 Not exhaustive list of the main calorimetric and thermal experimental methods

Name of the technique	Abbreviation	Measured quantity	Typical uses
Thermogravimetry	TG	Mass	Decompositions
Thermogravimetric analysis	TGA	Mass	Oxidations
Differential thermal analysis	DTC	Temperature difference	Phase changes, reaction analysis
Differential scanning calorimetry	DSC	Power difference and/or heat flow	Heat capacity, phase change, reaction analysis
Thermomechanical analysis	TMA	Deformation	Mechanical changes
Dielectric thermal analysis	DETA	Electrical response	Phase changes, phase transitions
Thermoptometry		Optical quantities	Phase changes, surface reaction analysis, colour changes
Thermoluminescence	TL	Light emitted	Oxidation
Thermomagnetometry	TM	Magnetic response	Magnetic changes and phases

spectrometric techniques. The involved thermodynamic changes often modify one or more of these material properties: the changes can be either physical, such as melting, crystalline and symmetry transitions, order-state transitions, etc., or they can be chemical, involving reactions altering the composition of the material.

2.2 Electric and Spectroscopic Methods

Among physical changes, the thermal evolution of the electromagnetic properties are often considered very important and meaningful, both for fundamental and applicative studies. Moreover, from the point of view of experimental techniques and apparatus, the spectroscopic techniques widely employed in matter physics laboratories usually exploit or investigate electromagnetic properties: this makes possible to overlap thermal and electric experimental analyses by starting either from usual spectroscopic or calorimetric setups, then upgrading them to also record further properties.

In this vein, a well-established spectroscopic technique is dielectric spectroscopy (DS), which over the last decades has expanded by largely expanding the available frequency region, leading to the development of highly sensitive *broadband dielectric spectroscopy* (BDS) [19]. Such a technique measure the response of a sample under an oscillating electric field; its dielectric features are measured as a function of frequency, in terms of *complex dielectric function*, $\varepsilon = \varepsilon' + i\varepsilon''$, whose both real ($\varepsilon'$) and imaginary ($\varepsilon''$) parts are usually obtained. An advantage of such method is that it can be used on a very large variety of compounds and samples, such as composite and nonlinear materials [20, 21], ionic liquids [22], molecular glass-forming systems [23], geometrically-confined systems [24], etc. Moreover, it can also be used to study transients and time evolution of samples, e.g. crystallization processes [25]. However, the specialty of this frequency-resolved experimental approach is the investigation of relaxation dynamics, for which it represents one of the most powerful choices [12, 19, 26]. This is the main reason why BDS (and the usual DS), exploits frequency as the independent variable. while temperature is usually treated as a parameter: indeed, a typical DS measurement is performed by spanning frequency at fixed temperature (see e.g. Fig. 1).

2.3 Matching Thermal and Electric Experimental Methods

On the other hand, one may adopt a different experimental perspective, by considering frequency as a parameter and temperature as the independent variable. Indeed, when the perturbing (static or oscillating) electric field is applied while the temperature is varied, the resulting experimental approach closely resembles that employed in thermal analyses. An example of such a kind of measurement is displayed by Fig. 2. In particular, this approach is akin to what is done with thermomechanical analysis (TMA) through the mechanical stress. Analogously to BDS techniques, in this kind of measurement the specimen is usually placed between two metal electrodes in order to obtain a geometric arrangement of parallel planes: this allows obtaining a simple plane-parallel-capacitor field configuration. In the field of thermal analysis technique, such approach is often known as *dielectric thermal analysis* (DETA) [15]; in other fields it can also be called temperature-resolved (or

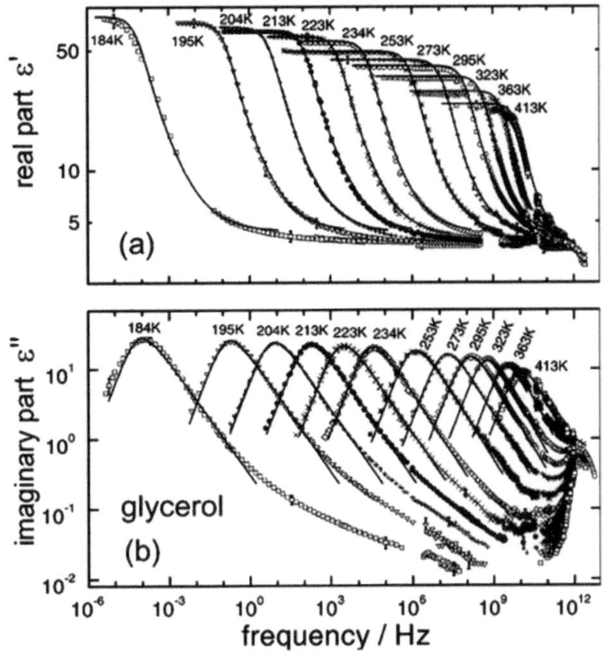

Fig. 1 Typical BDS measurement plot from Fig. 2.14 of [19], where frequency is spanned at several fixed temperature. Dielectric spectra: (**a**) real part ε'; (**b**) imaginary part ε'' for glycerol as measured in the entire frequency regime from 10^{-6} to 10^{12} Hz. The dynamic glass transition is seen over many decades in frequency with varying temperature. (Courtesy of the original publisher ©)

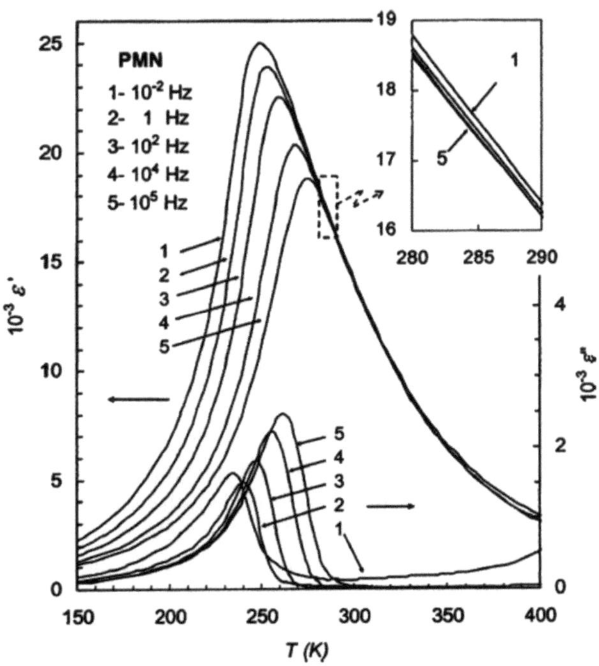

Fig. 2 Typical temperature-resolved dielectric measurement plot from Fig. 1 of [27], where temperature is spanned at several fixed frequencies. The plot shows the temperature dependences of the real (ε', left) and imaginary (ε'', right) parts of the relative dielectric permittivity measured at different frequencies in a crystal of the prototypical relaxor $Pb(Mg_{1/3}Nb_{2/3})O_3$. Enlarged view in the insert shows the so-called *universal relaxor dispersion*. (Courtesy of the original publisher ©)

temperature-dependent) dielectric measurement or temperature-resolved dielectric spectroscopy or similar ones [28–31].

In the practical implementation of this technique, the sample is usually subjected to an electric voltage V. Reliable measurements employing static fields are not technically feasible; therefore most of measurements are carried out with oscillating sinusoidal fields with angular frequency ω [15, 31]. While this polarizes the sample, the measurement apparatus measures a current flow I which has a dephasing δ with respect to the bias field. If we define the sample admittance Y as

$$Y = \frac{I}{V},$$

the corresponding capacitance C will be given by

$$C = \frac{|Y|}{\omega} \cos \delta,$$

which represents the ability to store electric charge and then electric energy. Similarly, the conductance of the sample is given by

$$G_c = |Y| \sin \delta.$$

Thus, the measurement procedure provides the capacitance C and the conductance G_c of the considered sample. From these quantities, the specific dielectric features of the material are obtained by eliminating the contribution due to the geometry of the specific sample. Hence, the real ε' and imaginary ε'' parts of the dielectric function are respectively given by

$$\varepsilon' = \frac{C}{\varepsilon_0 \Gamma} \tag{6}$$

and

$$\varepsilon'' = \frac{G_c}{\omega \varepsilon_0 \Gamma}, \tag{7}$$

where ε_0 is, as usually, the vacuum dielectric permittivity and Γ is a factor depending on the geometry of the sample. In the simplest and most used plane capacitor configuration, if A is the surface of the planes and D is the distance between them, so $\Gamma = A/D$. The ratio between ε'' and ε' allows to define an important quantity $\tan \delta = \varepsilon'/\varepsilon''$, or, more generally

$$\tan \delta \equiv \frac{\mathrm{Im}(\varepsilon)}{\mathrm{Re}(\varepsilon)}, \tag{8}$$

which is called *dissipation factor*, or *loss tangent*, and represents the amount of energy dissipated divided by the amount of energy stored, per cycle. It can be directly calculated from the measured capacitance and conductance, even without knowing the geometry of the system, so it is not affected by the errors associated with the measurement of involved sizes. It is a key quantity for the study of dissipation and relaxation phenomena [19, 32].

However, the ultimate goal of temperature-resolved dielectric measurements (or DETA) is to obtain the evolution of ε as a function of T. This is usually employed to detect phase transitions, Curie points, glass transitions, and similar phenomena, by interpreting the shape of the $\varepsilon = \varepsilon(T)$ curve. Nevertheless a more subtle approach can be employed. From a strictly thermodynamic and calorimetric point of view, we need to obtain an appropriate thermodynamic potential, e.g. some of the physical quantities listed in Table 1, that may be suitable to describe the considered system. Unlike standard calorimetry, where heat (and energy) can be directly measured, if we aim at exploiting dielectric measurements we must extract the thermodynamic quantities from the $\varepsilon = \varepsilon(T)$ curve. This can be achieved through specific mathematical relationships correlating the realm of thermodynamic potential with the realm of the material electric properties. To accomplish this goal, a theory should be developed, starting from reasonable assumptions, which should be minimal yet precise. This is precisely what is done by Herbert Fröhlich [33] in his seminal work, whose assumptions will be listed in the next section, while the specific formal steps will be exposed in the next chapter [6, 7]. Fröhlich's work is titled "Theory of dielectrics" and "is intended to give a systematic account of the theory of the dielectric constant and of dielectric loss."[3] Although based on a theoretical point of view, the monograph is intended "for the use of applied scientists".[4] This is shown by the general sound of the treatise and by the choice of the discussed topics. In fact, the first chapter explicitly presents the macroscopic (phenomenological) theory and distinguishes the cases of static and oscillating fields. At the end of the chapter Fröhlich exposes the classical macroscopic thermodynamic theory of ideal dielectric. Starting from general concepts on the thermodynamics of field energy in Maxwell's theory [34], he builds theory for the ideal case, that is relatively simple yet of very general validity.

Under suitable "light" hypotheses, Fröhlich's theory provides relationships (5) with explicit analytical functions. Therefore, the obtained equations can be used as the starting point to carry out quantitative calorimetric analyses through dielectric measurements. In particular, Fröhlich himself focused on the meaning of the *entropy contribution* (5b), to which he gave a relatively simple yet fruitful interpretation so that the field-dependent part $S_E(T, E)$ is referred to as the *Fröhlich entropy* (FE) [29, 35]. Both such mathematical tools and the physical interpretation, which we are discussing in the next section, paved the way of an original experimental approach in matter physics and material science, a method overlapping experimental

[3] H. Fröhlich in [7].

[4] ibid.

techniques which were previously considered quite far, the *Fröhlich entropy estimation* [35].

3 Framework and Basic Interpretation of Fröhlich Entropy

After the above mentioned theories on the thermodynamics of electromagnetic fields [34], and the successive studies the electrocaloric effects (Sect. 1) [1, 3], which were conducted mainly during the 1930s and 1940s, the development of a thermodynamics theory of dielectric materials was proposed in a few important works. These were done by prominent scientists such as the above recalled H. Fröhlich [6, 7], as well as R. Becker [8], L.D. Landau and E.M. Lifšic [9], V. Daniel [10], C. Böttcher [11], and B. Scaife [12]. The formulation by Fröhlich is, however, the cornerstone upon which the others were built, so we are here following his steps to obtain the basic equations of dielectrics thermodynamics. The mathematical details of the theory will be presented in chapter "Elements of Thermodynamics of Dielectrics and Basic Theory of Fröhlich Entropy". In particular, the relationship describing the field-induced entropy (the above recalled *Fröhlich entropy*), will be recalled together with the simple interpretation given by Fröhlich himself. This interpretation constitutes the foundation of the Fröhlich entropy estimation (FE estimation) thermodynamic method. A deeper and more subtle discussion about the physical meaning of FE will be given in the subsequent chapter "Meaning and Applications of Fröhlich Entropy".

3.1 Assumptions and Results of Fröhlich's Theory

We previously said that an explicit mathematical functional formulation of the (5) requires the introduction of some basic assumptions to start. Discussions on the assumptions of Fröhlich's theory have identified six basic hypotheses [35, 36]. The equations are aimed at providing the explicit analytical form of the (5). The six specific assumptions to be fulfilled are the following ones.

There are 2 *assumptions about the thermodynamics* of the considered process, i.e. isothermal conditions and reversibility. Moreover, 4 *assumptions about the* properties of the *dielectric* are taken into account. Two of them are strictly correlated to the thermodynamic behavior of the considered dielectric. They are negligible volume variation (no electrostrictions, etc.) and ideal dielectric (loss-free medium). Finally, the last 2 are correlated to the electromagnetic response of the material, so **D** is considered to linearly depend on **E** and the dielectric materials is considered isotropic (scalar approximation).

These last two items are not associated with specific physical limitations. They are instead assumed with the aim of simplifying the mathematical procedure, since it has been shown that Fröhlich's theory can be straightforwardly generalized to the case of a dielectric with anisotropic and nonlinear responses [36]. Such topic will be discussed in chapter "Generalized Fröhlich Entropy: Anisotropic and Nonlinear Media".

Finally, we summarize the details of the theory by stating that Fröhlich's results demonstrate that, under the above-stressed assumptions, relationships (5) assume the following mathematical form:

$$U(T, E) = U_0(T) + \frac{1}{2}\left(\varepsilon + T\frac{\partial \varepsilon}{\partial T}\right) E^2, \tag{9a}$$

$$S(T, E) = S_0(T) + \frac{1}{2}\frac{\partial \varepsilon}{\partial T} E^2, \tag{9b}$$

$$F(T, E) = F_0(T) + \frac{1}{2}\varepsilon E^2, \tag{9c}$$

which are the *Fröhlich's thermodynamic relationships* [7], i.e. respectively *Fröhlich total energy*, *Fröhlich entropy*, *Fröhlich free energy* (always for unitary volume) [29, 30, 37]. We stress that the (9) points out that the usual standard formula for energy density U_e, which is reported in every textbook on electromagnetism, i.e.

$$U_e = \frac{1}{2}\mathbf{D} \cdot \mathbf{E}, \tag{10}$$

does provide the Helmholtz free energy (F) and not the total internal energy U.

3.2 Comments on the Theory

Expressions (9) possess a clear and simple mathematical form: the free energy is demonstrated to be proportional to the dielectric linear response ε and the entropy is shown to be proportional to its temperature derivative. Neverthless, such an elegant formulation needs the above-recalled specific premises. It is worth discussing some of them in greater detail.

1. The use of the standard basic thermodynamic theory requires that only reversible transformations can be considered. This implies that, at each examined temperature value, the system should be in thermodynamic equilibrium, so that no time-dependent physical variable should be involved [18]. This means that, strictly speaking, Eqs. (5) are valid in static conditions only. In practice, this is satisfied when the investigated dielectric does not exhibit significant relaxation phenomena [38–41]. This provides a practical experimental criterion to discriminate whether Fröhlich's thermodynamic relationships can or cannot be used. They hold when the imaginary part $\varepsilon'' = \mathrm{Im}(\varepsilon)$ of the dielectric response ε is negligible. This happens when that dielectric loss tangent (8) is as small as possible [42]: in terms of standard dielectric spectroscopy it reads

$$|\tan \delta| \ll 1. \tag{11}$$

Given that no real dielectric materials is ideal, and therefore ε'' is never 0, the (11) provides a condition that can be reasonably and fruitfully exploited to use the (9) on real samples. In fact, a verification of the condition (11) is a prerequisite to apply Fröhlich's theory for the interpretation of real data.

2. Such equations, despite their apparent simplicity, are able to highlight interesting and important features. For example, Eq. (9c) for the free energy shows that the quantity $\varepsilon_0 \varepsilon_s E^2/2$ corresponds to the amount of electric energy available in an isothermal thermodynamic transformation. On the other hand, Eq. (9a) indicates that in the compounds which obeys to the Curie-Weiss law, e.g. diluted dipolar gases, the part of ε_s which depends on temperature does not contribute to the total energy of the system [7].[5]
3. In the present discussion about the dielectric calorimetry through *Fröhlich entropy estimation*, the key physical quantity among the (9) is the entropy:

$$S(T, E) = S_0(T) + S_E(T, E) = S_0(T) + \frac{1}{2}\frac{\partial \varepsilon}{\partial T} E^2,$$

whose the field dependent part is

$$S_E(T, E) = \frac{1}{2}\frac{\partial \varepsilon}{\partial T} E^2. \tag{12}$$

This formula pinpoints that *Fröhlich entropy* can be interpreted as the entropy variation of the system due to the application of the electric field E (for volume unit) [7]. For practical uses, it is useful having $S_E(T)$ for a unitary field [29, 30, 43, 44], i.e.

$$s(T) = \frac{1}{2}\frac{\partial \varepsilon}{\partial T}. \tag{13}$$

Since $s(T)$ does not depend on the strength of E, it highlights the information about the material given by the dielectric response. In fact, Eq. (13) is the *key relationship* for exploiting *FE evaluation technique* on the most of the physical systems [29, 30, 35, 43, 44].
4. Finally, the strength of E should be carefully considered: when very high fields are used (typically $\gtrsim 10^4$ V/cm), the investigated material can be strongly perturbed, so E could modify also the very structure of the sample, giving rise e.g. to coercive field effects [41, 42, 45, 46]. For this reason, the probing electric field typically employed in the experiments which employ Fröhlich's thermodynamic theory is typically $\lesssim 5$ V/cm.

[5] Pg. 12, § 3, Eqs. (11) and (14).

3.3 Fröhlich Entropy Interpretation and Practical Uses

Fröhlich entropy has a deep physical meaning, which goes far beyond the description of a dielectric material subjected to an electric field during a thermodynamic process. While a detailed discussion about the interpretation of FE entropy will be carried out in the following chapter "Meaning and Applications of Fröhlich Entropy", we here provide the basic elements to grasp its importance for the thermodynamic of dielectrics and for the development of the FE estimation experimental technique. We start by quoting Fröhlich himself from [7].[6]

> Finally, Eq. 12 [$S = S_0(T) + (\partial\varepsilon_s/\partial T)(E^2/8\pi)$, i.e. Eq. (9b) above, editor's note] shows that the entropy is increased by the field if $\partial\varepsilon_s/\partial T$ is positive, and decreased if this quantity is negative. Since the entropy is a measure of the molecular disorder, an external field creates order in dipolar liquids and gases for which ε_s decreases with increasing T. This may be expected because the field will orientate some of *the dipoles which in the absence of a field are at random*. In some dipolar solids, on the other hand ε_s increases with T, which means that an external field increases the disorder. *This too is understandable if one assumes that in the absence of a field the dipoles are in a well-ordered state as may be expected in solids.* The field by turning some of the dipoles into different directions can thus only decrease the existing order.[7]

Just after this statement Fröhlich's textbook displays a schematic plot analogous to that in Fig. 3, whose caption is

> Schematic temperature-dependence of the dielectric constant ε_s and of the entropy change $S \propto \partial\varepsilon_s/\partial T$ due to the polarization by a field. If $S > 0$ the field creates disorder, if $S < 0$

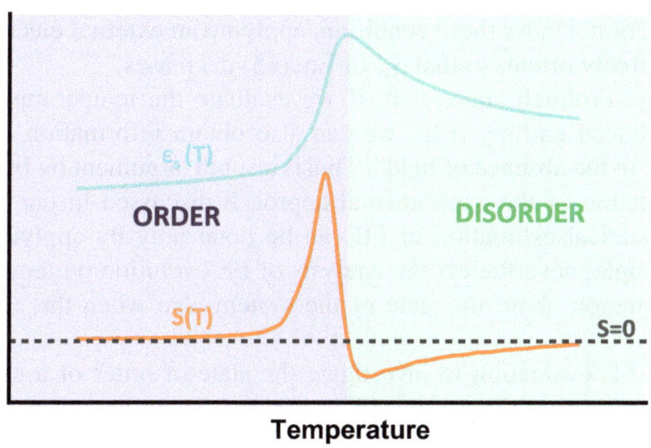

Fig. 3 Scheme of the dielectric constant ε_s on temperature (light blue curve) with its related the field-induced entropy change, following the ideal model reported by Eq. (12): $S_E \propto \partial\varepsilon_s/\partial T$ (orange curve). Fröhlich pointed out that, if $S_E > 0$, the field will produce disorder, if $S_E < 0$, it will produce order

[6] Chapter "Introduction: Dielectric Calorimetry and Fröhlich Entropy Estimation", pp. 12–13.

[7] Emphasis added.

it creates order. Near the absolute zero of temperature the substance is *already* perfectly ordered. Hence $\partial \varepsilon_s / \partial T$ cannot be negative near $T = 0$.[8]

Such Fröhlich's statement can be highlighted by means of a simple example, i.e. if we analyze an ideal physical system following the Curie-Weiss law. Let's consider a system of dipoles, displaying an idealized dielectric response such in Fig. 3. This Plot exhibits a peak temperature of the $\varepsilon_s = \varepsilon_s(T)$ curve, which can be identified as a transition temperature T_C. It marks the transition between two distinct behaviors. In the *first region*, below T_C, the considered dielectric system displays some kind of ordered structure, so that Curie-Weiss law can't be applied. Instead, Curie-Weiss law holds when the physical system is constituted by non-interacting randomly-ordered dipoles in a mean-field, whose orientation at zero field is due to statistical thermal distribution (paraelectric phase). So that, in the *second region*, for $T > T_C$, the (relative static) dielectric response is given by

$$\varepsilon_s(T) = \frac{C}{T - T_C}, \tag{14}$$

where C is called the *dielectric Curie-Weiss constant*[9] [42]. If we derive Eq. (14) to apply Fröhlich's relationship (12), we obtain

$$S_E = \frac{-C}{(T - T_C)^2} \frac{\varepsilon_0}{2} E^2. \tag{15}$$

Following Fröhlich's interpretation, the negative sign in the right-hand side of Eq. (15) highlights the temperature range where the internal dipolar structure has a random orientation. Under these condition, applying an external electric field make the dipoles to freely orient, so that S_E of Eq. (15) decreases.

In summary, Fröhlich states that, if we evaluate the temperature dependence of the field-induced entropy (FE), we can also obtain information on the system configuration "in the absence of field". This is a subtle argument by Fröhlich, which lays the foundation of the experimental approach discussed in our present book. While the numerical estimation of FE can be done only by applying an electric field to the sample, nevertheless the analysis of FE evolution on temperature gives detailed information about the state of the system also when the electric field is absent.

The use of FE evaluation to investigate the state-of-order of a system lays on this sharp interpretation. Therefore, FE estimation experimental approach can be summarized as following [35].

1. The dielectric response $\varepsilon = \varepsilon(T)$ of a given dielectric sample is suitably measured on temperature.

[8] Emphasis added.

[9] Of course $C > 0$.

Fig. 4 Gianbattista Parravicini (b. 1940) photographed in the Dielectric Laboratory at the University of Pavia's Department of Physics (Italy), where *FE estimation technique* was conceived in the early 2000s. (Courtesy of Prof. GB. Parravicini)

2. The satisfaction of Fröhlich's assumptions by the sample is confirmed.
3. The Fröhlich entropy is calculated from the data through the suitable Fröhlich relationships.
4. The data are analyzed following the interpretation of Fröhlich entropy.

In particular, thanks to Fröhlich's interpretation of S_E, the most significant feature of FE estimation lies in its ability to provide specific information about the state of order/disorder of the system, to describe and follow its evolution on temperature.

Although the validity of Fröhlich's thermodynamic equations (9) was recognized by the most influential subsequent theorists of dielectrics [8–12], no specific experiments had been carried out until 2003, when Gianbattista Parravicini (Fig. 4) et al. from University of Pavia (Italy) employed Fröhlich's thermodynamic theory, equations and interpretation to investigate melting processes in confined metallic nanoparticles [47–49]. These works employ both Eq. (9b) and the above-recalled original interpretation, so devising an innovative experimental approach and paving the way of the Fröhlich entropy estimation technique. Afterwards, the same concepts were successfully used to study liquid crystals and, in particular, nematic compounds [50–56], molecular crystals [57, 58], ferroelectric and glassy perovskites [29, 30, 37, 43, 59–61], polar glasses [62], polar and glass-forming liquids [45, 63–66], polymeric films [67], enzymes [68].

This brief landscape shows that *FE estimation technique* has been successfully applied to a wide variety of physical systems. It has been able to give results, especially regarding the analysis of the state-of-order, in full agreement with best-known experimental methods, whose we are discussing in chapter "Experimental Techniques to Investigate the State-of-Order in Condensed Matter". All these items represent a validation of such method, indicating that it can be a feasible and reliable technique, complementary to the other widely employed techniques of thermal analysis, calorimetry, structural analysis, etc.

References

1. J. Scott, Annu. Rev. Mater. Res. **41**, 229 (2011)
2. R. Newhman, *Properties of Materials - Anisotropy, Symmetry, Structure* (Oxford University Press, Oxford, 2005)
3. A. Devonshire, Philos. Mag. **40**, 1040 (1949)
4. A. Devonshire, Adv. Phys. **3**, 85 (1954)
5. J. Nye, *Physical Properties of Crystals - Their Representation by Tensors and Matrices* (Claredon Press, Oxford, 1985)
6. H. Fröhlich, *Theory of Dielectrics*, 1st edn. (Claredon Press, Oxford, 1949)
7. H. Fröhlich, *Theory of Dielectrics*, 2nd edn. (Claredon Press, Oxford, 1958)
8. R. Becker, *Electromagnetic Fields and Interactions* (Blackie and Son, London, 1964)
9. L. Landau, E. Lifshitz, *Electrodynamics of Continuous Media* (Pergamon Press, Oxford, 1960)
10. V. Daniel, *Dielectric Relaxation* (Academic Press, London, 1967)
11. C. Böttcher, *Theory of Electric Polarization*, 2nd edn. (Elsevier Science, 1973)
12. B. Scaife, *Principles of Dielectrics* (Claredon Press, Oxford, 1998)
13. S. Duckett, B. Gilbert, *Foundations of Spectroscopy* (Oxford University Press, Oxford, UK, 2000)
14. W. Demtröder, *Laser Spectroscopy* (Springer, Berlin, Heidelberg, Deutschland, 2013)
15. P. Haines (ed.), *Principles of Thermal Analysis and Calorimetry* (Royal Society of Chemistry, Cambridge, UK, 2002)
16. M. Brown, *Introduction to Thermal Analysis*, 2nd edn. (Kluwer Academic Press, New York, 2001)
17. K. Huang, *Statistical Mechanics* (John Wiley & Sons, New York, 1987)
18. H. Callen, *Thermodynamics and an Introduction to Thermostatistics* (John Wiley & Sons, New York, 1985)
19. F. Kremer, A. Schönhals (eds.), *Broadband Dielectric Spectroscopy* (Springer, Berlin, Heidelberg, 2003)
20. A. Schönhals, P. Szymoniak (eds.), *Dynamics of Composite Materials* (Springer, Berlin, Heidelberg, 2022)
21. R. Richert (ed.), *Nonlinear Dieletric Spectroscopy* (Springer, Berlin, Heidelberg, 2018)
22. M. Paluch (ed.), *Dielectric Properties of Ionic Liquids* (Springer, Berlin, Heidelberg, 2016)
23. G. Floudas, M. Paluch, A. Grzybowski, K. Ngai (eds.), *Molecular Dynamics of Glass-Forming Systems* (Springer, Berlin, Heidelberg, 2011)
24. F. Kremer (ed.), *Dynamics in Geometrical Confinement* (Springer, Berlin, Heidelberg, 2014)
25. T. Ezquerra, A. Nogales (eds.), *Crystallization as Studied by Broadband Dielectric Spectroscopy* (Springer, Berlin, Heidelberg, 2020)
26. F. Kremer, A. Loidl (eds.), *The Scaling of Relaxation Processes* (Springer, Berlin, Heidelberg, 2018)
27. A. Bokov, Z. Ye, Front. Ferroelectr. **41**, 31 (2006)
28. L. Dissado, *Dielectric Response* (Springer International Publishing, Cham, 2017), 1 pp.
29. J. Parravicini, E. DelRe, A. Agranat, G. Parravicini, Phys. Rev. B **93**, 094203 (2016)
30. J. Parravicini, E. DelRe, A. Agranat, G. Parravicini, Nanoscale **9**, 9572 (2017)
31. V. Kochervinskii, I. Malyshkina, in *Organic Ferroelectric Materials and Applications*, ed. by K. Asadi, Woodhead Publishing Series in Electronic and Optical Materials (Woodhead Publishing, 2022), pp. 263–355
32. R. Horansky, L. Clarke, J. Price, T. Khuong, P. Jarowski, M. Garcia-Garibay, Phys. Rev. B **72**, 014302 (2005)
33. N. Mott, Biograph. Mem. Fellows R. Soc. **38**, 147 (1992)
34. M. Abraham, R. Becker, *The Classical Theory of Electricity and Magnetism* (Blackie & Son Limited, London, 1932)
35. J. Parravicini, G. Parravicini, Res. Phys. **28**, 104571 (2021)
36. J. Parravicini, Phys. B **541**, 54 (2018)

37. P. Tan, H. Tian, F. Huang, X. Meng, Y. Wang, C. Hu, X. Cao, L. Li, Z. Zhou, Phys. Rev. Appl. **11**, 024037 (2019)
38. W.T. Coffey, Y.P. Kalmykov (eds.), *Fractals, Diffusion, and Relaxation in Disordered Complex Systems: Advances in Chemical Physics, Part B* (John Wiley & Sons, Hoboken, 2006)
39. S.J. Rzoska, V.P. Zhelezny (eds.), *Nonlinear Dielectric Phenomena in Complex Liquids* (Kluwer Academic Publisher, New York, 2004)
40. P. Lunkenheimer, M. Koehler, S. Kastner, A. Loidl, *Dielectric Spectroscopy of Glassy Dynamics* (John Wiley & Sons, Hoboken, 2012), chap. 3, pp. 115–149. https://doi.org/10.1002/9781118202470.ch3
41. P. Lunkenheimer, F. Humann, D. Reuter, K. Geirhos, A. Loidl, G. Johari, Phys. Rev. Mater. **3**, 112601 (2019)
42. K. Kao, *Dielectric Phenomena in Solids - With Emphasis on Physical Concepts of Electronic Processes* (Elsevier Academic Press, San Diego, 2004)
43. F. Huang, C. Hu, H. Tian, X. Meng, P. Tan, Z. Zhou, CrystEngComm **21**, 7002 (2019)
44. F. Huang, C. Hu, Z. Zhou, X. Meng, P. Tan, Y. Wang, C. Wang, H. Tian, Acta Mater. **200**, 24 (2020)
45. G. Johari, J. Chem. Phys. **138**, 154503 (2013)
46. F. Jona, G. Shirane, *Ferroelectric Crystals* (Dover, New York, 1993)
47. G. Parravicini, A. Stella, P. Tognini, P. Merli, A. Migliori, P. Cheyssac, R. Kofman, Appl. Phys. Lett. **82**, 1461 (2003)
48. G. Parravicini, A. Stella, M. Ungureanu, P. Merli, A. Migliori, P. Cheyssac, R. Kofman, Eur. Phys. J. D **24**, 219 (2003)
49. G. Parravicini, A. Stella, M. Ungureanu, P. Merli, A. Migliori, R. Kofman, Phys. Status Solidi B **237**, 374 (2003)
50. E. Thoms, L. Yu, R. Richert, J. Mol. Liquids **368**, 120664 (2022)
51. J. Jadżyn, G. Czechowski, J. Phys. Chem. B **111**, 3727 (2007)
52. J. Jadżyn, U. Sokołowska, J. Déjardin, J. Phys. Chem. B **112**, 9050 (2008)
53. J. Jadżyn, G. Czechowski, J. Déjardin, J. Phys. Chem. B **112**, 4948 (2008)
54. J. Jadżyn, J. Swiergiel, D. Bauman, R. Dąbrowski, K. Czupryński, U. Sokołowska, Phase Transit. **82**, 409 (2009)
55. D. Bauman, J. Jadżyn, E. Wolarz, A. Modlińska, R. Dąbrowski, Opto-Electron. Rev. **18**, 63 (2010)
56. J. Świergiel, I. Płowaś, J. Jadżyn, J. Mol. Liq. **220**, 879 (2016)
57. A. Sassella, D. Braga, M. Campione, T. Ciabattoni, M. Moret, J. Parravicini, G. Parravicini, J. Appl. Phys. **109**, 013529 (2011)
58. G. Parravicini, M. Campione, F. Marabelli, M. Moret, A. Sassella, Phys. Rev. B **86**, 024107 (2012)
59. L. Lo Presti, J. Parravicini, R. Soave, G.B. Parravicini, M. Mauri, L. Loconte, F.D. Mei, L. Falsi, L. Tartara, S. Binetti, A.J. Agranat, E. DelRe, Phys. Rev. B **102**, 214110 (2020)
60. E. DelRe, F.D. Mei, J. Parravicini, G. Parravicini, A. Agranat, C. Conti, Nat. Photon. **9**, 228 (2015)
61. J. Parravicini, E. DelRe, S. Perego, M. Acciarri, S. Binetti, Y. Garcia, G. Parapelitsa, A. Agranat, G. Parravicini, Phys. Rev. B **106**, 064107 (2022)
62. D.V. Matyushov, J. Chem. Phys. **145**, 034504 (2016)
63. G. Johari, J. Chem. Phys. **145**, 164502 (2016)
64. G. Johari, J. Chem. Phys. **139**, 026101 (2013)
65. S. Samanta, R. Richert, J. Chem. Phys. **142**, 044504 (2015)
66. S. Samanta, R. Richert, J. Phys. Chem. B **120**, 7737 (2016)
67. V.V. Kochervinskii, S.A. Bedin, I.V. Razumovskaya, M.A. Shcherbina, M.I. Buzin, I.A. Malyshkina, A.M. Prazdnichnyi, A. Pavlov, O. Vasilevskii, Polym. Sci. Ser. A **58**, 345 (2016)
68. Y. Kurzweil-Segev, A.G. (Gutina), I. Popov, D. Golodnitsky, Y. Feldman, Phys. Chem. Chem. Phys. **18**, 10992 (2016)

Elements of Thermodynamics of Dielectrics and Basic Theory of Fröhlich Entropy

Jacopo Parravicini, Martin Tress, and Friedrich Kremer

Abstract This chapter provides the basic elements of thermodynamics of linear dielectrics. It begins by examining the fundamental thermodynamic relationships of dielectric materials under static electric fields, including definitions of thermodynamic potentials like energy, entropy, Helmholtz free energy, and how these quantities depend on the applied field. The categorization of macroscopic temperature-dependent effects and quantities related to dielectric behavior are discussed. Starting from ideal, loss-free dielectrics in equilibrium, the chapter outlines the formal construction of the thermodynamic potentials within H. Fröhlich's theory. It then addresses key applications such as phase transitions in ferroelectrics, polar liquids, and nematogenic systems. Finally, the thermodynamic response of non-ideal dielectrics is described in broad terms from a phenomenological perspective, highlighting the role of relaxation processes and energy dissipation in real materials.

1 General Framework

Elementary electromagnetic theory tells us that an electric field applied to a dielectric material causes polarization. In fact, this polarization arises from mechanisms such as electronic polarization, when electrons and atomic nuclei shift in opposite directions, ionic polarization, involving opposite movements of positive

J. Parravicini (✉)
Dipartimento di Fisica & Astronomia, Università degli Studi di Firenze, Sesto F., Firenze, Italy
e-mail: jacopo.parravicini@unifi.it

M. Tress
Molecular Photonics group, Peter Debye Institute for Soft Matter Physics, Faculty of Physics and Earth System Sciences, Universität Leipzig, Leipzig, Germany
e-mail: martin.tress@uni-leipzig.de

F. Kremer
Peter Debye Institute for Soft Matter Physics, Faculty of Physics and Earth System Sciences, Universität Leipzig, Leipzig, Germany
e-mail: friedrich.kremer@physik.uni-leipzig.de

J. Parravicini (ed.), *Fröhlich Entropy Estimation and Dielectric Calorimetry*,
Advances in Dielectrics, https://doi.org/10.1007/978-3-032-21454-6_2

and negative ions, and orientational polarization, due to the presence of permanent dipoles, e.g. ions or molecules, which align with the field. From a mathematical point of view, such dielectric features of a material are described by means of the (di)electric susceptibility function [1–3]. This mathematical object describes how the atoms of the material become polarized in practice.

Indeed, because the states of electrons and molecules change with temperature, the permittivity is also temperature-sensitive. This is why the dielectric constant, usually indicated by ε,[1] isn't "constant" at all: it includes both a static component and a component that varies with frequency, in addition to its sensitivity to temperature and pressure. In particular, the frequency-dependent behavior of the dielectric constant, arising from electronic, ionic, and dipolar contributions, is a large and well-established topic in condensed matter physics, which is specifically addressed elsewhere [3, 4]. Instead, this chapter focuses on the most basic and general relationships and effects in dielectrics subjected to thermodynamic changes, so that detailed discussions about relaxation phenomena, arising when the dependence of the studied phenomena on frequency is considered, will be avoided.

In the present chapter, we are providing a general overview about the thermodynamics of linear dielectrics, as the supporting framework of Fröhlich's entropy theory and interpretation. First, we consider the case of ideal loss-free dielectric material, for which we provide the basic thermodynamic relationships in the context of standard equilibrium thermodynamics, thus providing the formal construction of Fröhlich's theory to obtain thermodynamic potentials [5–11]. Then, we briefly recall some important thermodynamic phenomena in these materials, correlated with their macroscopic states, phases, and phase transitions [2, 12–15]. Finally, we briefly discuss the main thermodynamic effects in non-ideal dielectrics, i.e. in the presence of relaxation, where the behavior of their correlated thermodynamic physical quantities will be evaluated from a phenomenological perspective [1, 16–25].

2 Basic Thermodynamics of Dielectrics: Ideal Dielectrics

2.1 Key Concepts

Dielectric theory describes the polarization response of a dielectric material when it is subjected to an electric field. On the other hand, thermodynamics describes the response of a physical system subjected to a variation of temperature, heat, or similar parameter. If we combine these considerations, we may see that the concept of "thermodynamics of dielectrics" is an intrinsically hybrid topic, where the response of a physical system is simultaneously investigated with respect to both thermodynamic and electrical variations. Here we give the most general ideas

[1] In some texts is also indicated with κ.

about thermodynamics of dielectrics, in agreement with the conditions under which Fröhlich's entropy theory and concepts have been applied so far. We therefore consider:

1. linear dielectrics ($\mathbf{D} \propto \mathbf{E}$);
2. standard equilibrium thermodynamics.

It is worth noting that the condition of dealing with equilibrium processes implies that no time-dependent phenomena are involved in the theory we are recalling below. This also implies that ideal dielectric responses are considered, i.e. the imaginary part of the dielectric permittivity is negligible. In contrast, a comprehensive thermodynamic theory of dielectric in presence of the relaxation phenomena would require situating the exposed model in the framework of non-equilibrium thermodynamic theory of polarized systems, a condition beyond Fröhlich's hypotheses, which is beyond of the scope of this exposition [26, 27]. Some aspects about thermodynamic behavior in the presence of relaxation processes will be recalled in Sect. 5 from a strictly phenomenological point of view, without introducing or employing the tools of a general theory.

2.2 *Constitutive Relationships*

When a sample receive some external stimulus, such excitation can be taken into account by physical quantities, which are mathematically independent variables, while the thermodynamics response of the system is described through some suitable thermodynamic potential, the most important of which is the *energy*.[2] The amount of energy embedded in a physical system may be modified in several ways: usually, the change of the internal energy amount in a generic material can have thermal, electric,[3] or mechanical sources. Therefore, under the above-recalled hypotheses, i.e. when a reversible thermodynamics process is considered, an infinitesimal change in *temperature*, dT, generates an *entropy* change

$$dS = \frac{C}{T} dT, \tag{1}$$

where C is the *heat capacity*. A change of external *electric field* $d\mathbf{E}$ determines a change of *electric displacement*

$$d\mathbf{D} = \boldsymbol{\varepsilon} \cdot d\mathbf{E}, \tag{2}$$

[2] See Table 1 in chapter "Introduction: Dielectric Calorimetry and Fröhlich Entropy Estimation".

[3] More generally, it can be *electromagnetic*; however, we here consider cases of negligible magnetic responses.

where ε is the usual dielectric constant. Similarly, when there is a small change of *stress* (tensor) $d\boldsymbol{\sigma}$ is applied, the corresponding change of *strain* (tensor) will be

$$d\boldsymbol{\eta} = \boldsymbol{\Sigma} \cdot d\boldsymbol{\sigma}, \tag{3}$$

where $\boldsymbol{\Sigma}$ is the *elasticity tensor*.[4] We note that, in systems with high degree of symmetry, where $\boldsymbol{\sigma}$ cannot be uniquely defined, the mechanical contribution can be taken into account by substituting it with pressure p and volume v, so that (3) is replaced by

$$dv = \kappa_T dp, \tag{4}$$

where κ_T is often called (isothermal) *compressibility*.

Relationships (1), (2), and (3) allow us to discriminate independent (T, $\mathbf{E}$, $\boldsymbol{\sigma}$) from dependent ($S, \mathbf{D}, \boldsymbol{\eta}$) variables. We may use them as basic ingredients to develop a state function of the system. We first consider total energy U, whose differential, by applying the first principle of thermodynamics, is

$$dU = đQ - đW, \tag{5}$$

where, as usual, Q is the exchanged heat and W is the mechanical work. For a *reversible* process we have

$$d\,Q = TdS.$$

Therefore, we may combine these relationships with (1), (2), and (3), so (5) becomes

$$dU = TdS + \mathbf{E} \cdot d\mathbf{D} + \boldsymbol{\sigma} \cdot d\boldsymbol{\eta}. \tag{6}$$

Instead of (6), which is inconvenient for developing calculations, it is preferred to use Gibbs free energy[5] $G = G(T, \mathbf{E}, \boldsymbol{\sigma})$, which is used to derive relationships between the considered physical quantities. This reflects the actual conditions of most experimental configurations where temperature, electric field, stress $\boldsymbol{\sigma}$ are the independent variables and S, $\mathbf{D}$, $\boldsymbol{\eta}$ are the dependent ones [8, 9]. In the reversible transformations considered here, we will have

$$G = -ST - \mathbf{D} \cdot \mathbf{E} - \boldsymbol{\eta} \cdot \boldsymbol{\sigma} \tag{7}$$

---

[4] Also called *elastic compliance*.

[5] See Table 1, chapter "Introduction: Dielectric Calorimetry and Fröhlich Entropy Estimation".

whose the differential will have the form

$$dG = -S\,dT - \mathbf{D}\cdot d\mathbf{E} - \boldsymbol{\eta}\cdot d\boldsymbol{\sigma}. \tag{8}$$

From (8) we may obtain

$$S = -\left(\frac{\partial G}{\partial T}\right)_{\mathbf{E},\boldsymbol{\sigma}} \tag{9a}$$

$$\mathbf{D} = -\left(\frac{\partial G}{\partial \mathbf{E}}\right)_{T,\boldsymbol{\sigma}} \tag{9b}$$

$$\boldsymbol{\eta} = \left(\frac{\partial G}{\partial \boldsymbol{\sigma}}\right)_{T,\mathbf{E}}. \tag{9c}$$

Such expressions highlight the interrelations among entropy S, electric displacement $\mathbf{D}$, and strain $\boldsymbol{\eta}$, so that they can be derived through suitable differentiation. These also pinpoint the connections between piezoelectric and pyroelectric effects, as well as thermal expansion coefficients. Similar relations can be obtained by replacing $\boldsymbol{\sigma}$ with pressure p.[6]

2.3 *Maxwell's Relations for Dielectrics*

Several dielectric-thermodynamic physical constants are determined and mutually connected by calculating the second-order derivatives of (7); the obtained relations, which are the *Maxwell's thermodynamic relationships*, are equivalent to the first-order derivatives of (9). By reversing the order of differentiation, one obtains the matching coefficients for mutually-related phenomena. The following are the most widely used [12, 28, 29]. The specific heat (also called *heat capacity*):

$$-\left(\frac{\partial^2 G}{\partial T^2}\right)_{\mathbf{E}} = \left(\frac{\partial S}{\partial T}\right)_{\mathbf{E}} = \frac{C}{T}; \tag{10}$$

the *dielectric constant* (or *permittivity*):

$$-\left(\frac{\partial^2 G}{\partial E_i \partial E_j}\right)_{T,\boldsymbol{\sigma}} = \left(\frac{\partial \mathbf{D}}{\partial \mathbf{E}}\right)_{T,\boldsymbol{\sigma}} = \boldsymbol{\varepsilon}; \tag{11}$$

6 Relationship (8) becomes $dG = -S\,dT - \mathbf{D}\cdot d\mathbf{E} - v\,dp$, so that (9c) is $v = \left(\frac{\partial G}{\partial p}\right)_{T,\mathbf{E}}$.

the *elasticity* (or *elastic compliance*):

$$-\left(\frac{\partial^2 G}{\partial\sigma_{ij}\partial\sigma_{kl}}\right)_{T,\mathbf{E}} = \left(\frac{\partial\boldsymbol{\eta}}{\partial\boldsymbol{\sigma}}\right)_{T,\mathbf{E}} = \boldsymbol{\Sigma}; \tag{12}$$

the *direct piezoelectric coefficient*:

$$-\left(\frac{\partial^2 G}{\partial\sigma_{ij}\partial E_k}\right)_{T} = \left(\frac{\partial\mathbf{D}}{\partial\boldsymbol{\sigma}}\right)_{\mathbf{E},T} = \mathbf{d}, \tag{13}$$

which coincides with the *converse piezoelectric coefficient*, i.e.

$$-\left(\frac{\partial^2 G}{\partial\sigma_{ij}\partial E_k}\right)_{T} = \left(\frac{\partial\boldsymbol{\eta}}{\partial\mathbf{E}}\right)_{\sigma,T} = \mathbf{d}. \tag{14}$$

We may also obtain the *pyroelectric coefficient*

$$-\left(\frac{\partial^2 G}{\partial T\partial\mathbf{E}}\right)_{\boldsymbol{\sigma}} = \left(\frac{\partial\mathbf{D}}{\partial T}\right)_{\mathbf{E},\boldsymbol{\sigma}} = \left(\frac{\partial S}{\partial\mathbf{E}}\right)_{T,\boldsymbol{\sigma}} = \boldsymbol{\Pi}, \tag{15}$$

and the *thermal expansion coefficient* (or *piezocaloric effect coefficient*)

$$-\left(\frac{\partial^2 G}{\partial\boldsymbol{\sigma}\partial T}\right)_{\mathbf{E}} = \left(\frac{\partial\boldsymbol{\eta}}{\partial T}\right)_{\boldsymbol{\sigma},\mathbf{E}} = \left(\frac{\partial S}{\partial\boldsymbol{\sigma}}\right)_{T,\mathbf{E}} = \boldsymbol{\alpha} \tag{16}$$

All the above relations highlight the "coupling" among different physical variables determining the energy of the physical system. Such a "coupling" of physical quantities is at the heart of the concept of the "thermodynamics of dielectrics": an immediate and clear idea of such "coupling" of variables can be visualized and grasped through the so-called Heckmann diagram [8, 9, 30]. This is a graphical scheme originally developed to describe the macroscopic response of crystals, in order to highlight the various correlations among the involved quantities. There are several possible configurations of such a scheme, displaying different quantities arranged in different geometrical dispositions [8, 31], whose vertices represent external excitations of various nature (e.g. thermal, electrical, mechanical, etc.). The most usual configuration is a double triangle, one inside the other, such as in Fig. 1. Such a framework is usually adopted when the considered system exhibits a linear response. The *external triangle vertices* represent the *external stimulus* to the considered system i.e. the free variables *electric field* $\mathbf{E}$, *temperature* T, and *stress tensor* $\boldsymbol{\sigma}$. The *internal triangle vertices* represent the corresponding variables representing the response of the investigated system, i.e. the physical quantities depending on the free variables, which are *electric displacement* $\mathbf{D}$ *thermodynamic entropy* S, *strain tensor* $\boldsymbol{\eta}$. All these quantities can be combined to provide different forms of energy, which can be thermal, electrical, or mechanical. Each side

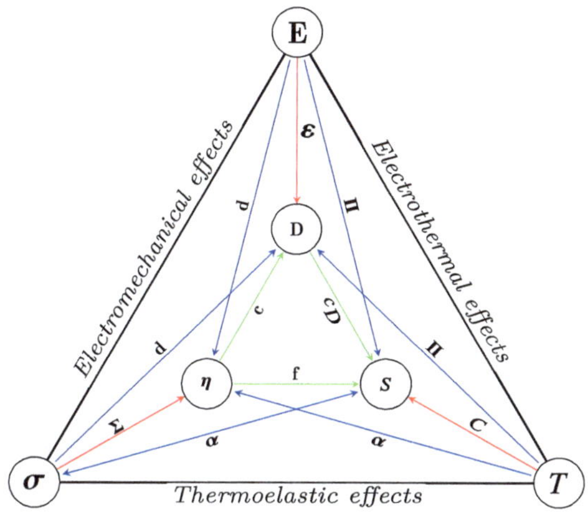

Fig. 1 Heckmann's diagram for crystalline materials, correlating electric fields, temperature, and mechanical stress fields [9, 31]. The *external vertices* display the main involved physical quantities, representing the external stimulus to the investigated system, i.e. the free variables, which are *electric field* **E**, *temperature* T, and *stress tensor* $\boldsymbol{\sigma}$. The *internal vertices* host the correlated variables representing the response of the considered system, i.e. the physical quantities depending on the free ones; these are *electric displacement* **D** *thermodynamic entropy* S, *strain tensor* $\boldsymbol{\eta}$. *Radial lines* (red) show the quantities correlating external and internal circle, i.e. *dielectric constant* $\boldsymbol{\varepsilon}$, *elasticity* $\boldsymbol{\Sigma}$, *heat capacity* C. *Among vertices* of the inner triangle, lines (green) indicate quantities correlating the response variables, which are *heat of polarization* c_D, *heat of deformation coefficient* **f**, *electrostriction coefficient* **c**. *Crossed lines* (blue) indicate physical quantities correlating inner and external vertices, which are *pyroelectric coefficient*, $\boldsymbol{\Pi}$, *thermal expansion coefficient* $\boldsymbol{\alpha}$, *direct piezoelectric effect coefficient* **d**, *converse piezoelectric effect coefficient* **d**. The exact definition of such quantities is listed in Table 1

describes a class of macroscopic phenomena: the $\boldsymbol{\sigma} - T$ side describes *thermoelastic* effects, the $T - \mathbf{E}$ side *electrothermal* ones, the $\mathbf{E} - \boldsymbol{\sigma}$ side *electromechanical* ones. Among the basic variables displayed at the vertices, several physical quantities can be formally defined (Table 1) and experimentally measured. Such variables describe the coupling of thermal, electric and mechanical variables. Some of them are calculated by expressing electric quantities as a function of thermal quantities, and vice-versa: these are the starting ingredient for a mathematical macroscopic description of dielectrics. Moreover, by exploiting the basic quantities shown by the diagram, we may obtain the total or the free energy, which can be used as state functions of the system.

Such a picture holds when the considered system is in a crystalline, or at least solid, configuration, namely where specific symmetries can be identified, so that their tensorial representation may be well defined [2, 9, 30]. However, a more generalized form of Heckmann diagram may be constructed by replacing the tensorial quantities, where symmetries play a key role, with more general mechanical quantities. In particular, pressure and volume may replace both stress and strain tensors, and they can also be easily defined in physical systems such

Table 1 Explanation of the physical quantities correlating the vertices of the triangles in Heckmann's diagram displayed in Fig. 1. T, $\mathbf{E}$, $\boldsymbol{\sigma}$ are *temperature*, *electric field*, and *stress tensor* respectively, while S, $\mathbf{D}$, $\boldsymbol{\eta}$ are *(total) entropy*, *electric displacement*, and *strain tensor* respectively. Several identities among these quantities are obtained through standard Maxwell's thermodynamic relationships [28]

First variable	Second variable	Quantity symbol	Quantity name	Formula
Radial lines (red)				
T	S	C	*Heat capacity* (or *specific heat*)	Eq. (10)
$\mathbf{E}$	$\mathbf{D}$	$\boldsymbol{\varepsilon}$	*Dielectric constant*	Eq. (11)
$\boldsymbol{\sigma}$	$\boldsymbol{\eta}$	$\boldsymbol{\Sigma}$	*Elasticity* (or *elastic compliance*)	Eq. (12)
Inner triangle (green)				
$\mathbf{D}$	S	c_D	*Heat of polarization*	$\left(\frac{\partial S}{\partial \mathbf{D}}\right) = c_D$
S	$\boldsymbol{\eta}$	$\mathbf{f}$	*Heat of deformation coefficient*	$\left(\frac{\partial \boldsymbol{\eta}}{\partial S}\right) = \mathbf{f}$
$\boldsymbol{\eta}$	$\mathbf{D}$	$\mathbf{c}$	*Electrostriction coefficient*	$\left(\frac{\partial \mathbf{D}}{\partial \boldsymbol{\eta}}\right) = \mathbf{c}$
Crossed lines				
T	$\mathbf{D}$	$\boldsymbol{\Pi}$	*Pyroelectric coefficient*	Eq. (15)
T	$\boldsymbol{\eta}$	$\boldsymbol{\alpha}$	*Thermal expansion coefficient*	Eq. (16)
$\boldsymbol{\sigma}$	$\mathbf{D}$	$\mathbf{d}$	*Direct piezoelectric effect coefficient*	Eq. (13)
$\mathbf{E}$	$\boldsymbol{\eta}$	$\mathbf{d}$	*Converse piezoelectric effect coefficient*	Eq. (14)

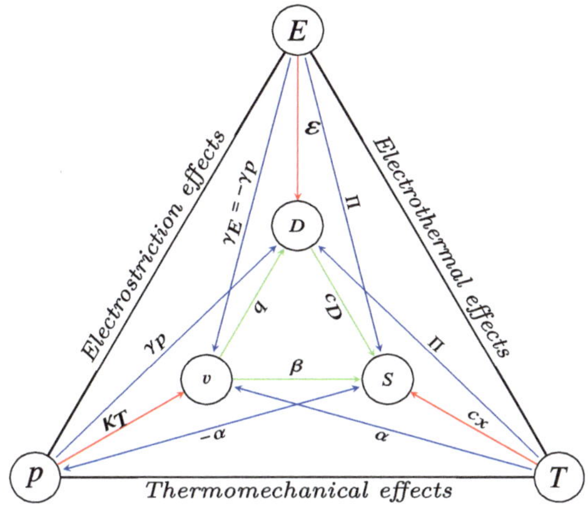

Fig. 2 Heckmann's diagram for more general classes of materials, with pressure and volume substituting stress and strain, and fields indicated in scalar approximation [9, 28, 31]. The *external vertices* display the main involved physical quantities, representing the external stimulus to the considered system i.e. the free variables *electric field* E, *temperature* T, and *pressure* p. The *internal vertices* host the correlated variables representing the response of the considered system, i.e. the physical quantities depending on the free ones; these are *electric displacement* **D** *thermodynamic entropy* S, *volume* v. *Radial lines* (red) show the physical quantities correlating external and internal circle, i.e. *dielectric constant* ε, *isothermal compressibility* κ_T, *heat capacity* C. *Among vertices* of the inner triangle, lines (green) indicate physical quantities correlating the response variables are *heat of polarization* c_D, *thermomechanical coefficient* β, *electrostriction coefficient* q. *Crossed lines* (blue) indicate physical quantities correlating inner and external vertices, which are *pyroelectric coefficient*, Π, *thermal expansion coefficient* α, *pressure coefficient of polarization* γ_p, *volumetric electrostriction* γ_E. Analogously to the diagram in Fig. 1, several identities may be obtained through Maxwell's relationships [28]

as liquids and gases. A possible generalized diagram is shown in Fig. 2, with the correlated physical quantities listed in Table 2. Here, the $p - T$ side describes *thermomechanical* effects, the $T - \mathbf{E}$ side the *electrothermal* ones, the $\mathbf{E} - p$ side the *electrostriction* ones.

3 Fröhlich's Theory of Dielectrics and Entropy

Beyond the very basic relationships obtained in Sect. 2, we here consider the development of a more detailed theory about the thermodynamics of dielectrics. The foundational thermodynamic theory for dielectric materials was proposed in several key works; such constructions were built on the earlier theories concerning the thermodynamics of electromagnetic fields and on subsequent studies of electrocaloric effects, which were recalled in the previous chapter (Sect. 1 in chapter "Introduction: Dielectric Calorimetry and Fröhlich Entropy Estimation") [5, 6, 32], that took

Table 2 Explanation of the physical quantities correlating the vertices of the triangles in generalized Heckmann's diagram displayed in Fig. 2. T, $\mathbf{E}$, p are *temperature*, *electric field*, and *pressure* respectively, while S, $\mathbf{D}$, v are *(total) entropy*, *electric displacement*, and *volume* respectively. Several identities among these quantities are obtained through standard Maxwell's thermodynamic relationships [28]

First variable	Second variable	Quantity symbol	Quantity name	Formula
Radial lines (red)				
T	S	C	*Heat capacity* (or *specific heat*)	Eq. (10)
$\mathbf{E}$	$\mathbf{D}$	ε	*Dielectric constant*	Eq. (11)
p	v	κ_T	*isothermal compressibility*	$\left(\frac{\partial v}{\partial p}\right)_{T,E} = -v\kappa_T$
Inner triangle (green)				
$\mathbf{D}$	S	c_D	*Heat of polarization*	$\left(\frac{\partial S}{\partial \mathbf{D}}\right) = c_D$
S	$\boldsymbol{\eta}$	β	*Thermomechanical coefficient*	$\frac{\partial v}{\partial S} = v\beta$
v	$\mathbf{D}$	q	*Electrostriction coefficient*	$\frac{\partial D}{\partial v} = q$
Crossed lines (blue)				
T	$\mathbf{D}$	Π	*Pyroelectric coefficient*	Eq. (15)
T	v	α	*Thermal expansion coefficient*	$\left(\frac{\partial v}{\partial T}\right)_{p,E} = -\left(\frac{\partial S}{\partial p}\right)_{T,E} = \alpha$
p	$\mathbf{D}$	γ_p	*Pressure coefficient of polarization*	$\left(\frac{\partial D}{\partial p}\right)_{E,T} = \gamma_p = -\gamma_E$
$\mathbf{E}$	v	γ_E	*Volumetric electrostriction*	$\left(\frac{\partial v}{\partial E}\right)_{p,T} = \gamma_E = -\gamma_p$

place primarily in the 1930s and 1940s. The contributions to the development of a thermodynamic theory of dielectrics came from seminal works by influential scientists, i.e. *Theory of dielectrics* by H. Fröhlich [10, 11], *Electromagnetic fields and interactions* by R. Becker [7], the well-known *Electrodynamics of continuous media* by L.D. Landau and E.M. Lifšic [33], *Dielectric relaxation* by V. Daniel [34], *Theory of dielectric polarization* by C. Böttcher [35], and *Principles of dielectrics* by B. Scaife [36]. Nevertheless, Fröhlich's formulation is widely considered the foundational work.[7] We will therefore follow his steps to derive the core equations of dielectrics thermodynamics. After establishing the relationship that defines the field-induced entropy variation (the previously mentioned *Fröhlich entropy*), we will present the straightforward interpretation provided by Fröhlich himself.

3.1 Basic Assumptions

In order to develop a mathematical formulation of the thermodynamic relationships of a dielectric, we start by recalling the general relationships (5) previously indicated in the chapter "Introduction: Dielectric Calorimetry and Fröhlich Entropy Estimation", providing the total energy U, the entropy S, and the Helmholtz free energy F for a dielectric subjected to a static electric field, i.e.

$$U(T, E) = U_0(T) + U_E(T, E), \tag{17a}$$

$$S(T, E) = S_0(T) + S_E(T, E), \tag{17b}$$

$$F(T, E) = F_0(T) + F_E(T, E), \tag{17c}$$

where U_0, S_0, F_0 are respectively the variation of total energy, entropy, Helmholtz free energy at zero field and U_E, S_E, F_E are their respective field-dependent parts. In order to provide an analytical expression of Eqs. (17) as a function of the dielectric response we need to establish a set of basic starting assumptions. Fröhlich's theory identifies 6 primary hypotheses [38, 39]. The specific purpose of these resulting equations is to furnish an analytical expression that fully describes the functional form of Relationships (17). These assumptions to be fulfilled are gathered in Table 3. Two of them are assumptions about the thermodynamics of the considered process, two are correlated to the thermodynamic behavior of the considered material, while

Table 3 Assumptions of Fröhlich's theory

Thermodynamic process	Behavior of material	Electromagnetic response
Reversible	Negligible volume variations	Linearity between **D** and **E**
Isothermal	Negligible losses	Scalar dependence of **D** on **E**

[7] It is worth noting that the basic Fröhlich's thermodynamic Relationships are employed also in the context of information theory, see. e.g. Ch. 9 of [37].

the last two are correlated to the electromagnetic response. These last hypotheses are assumed in order to simplify the treatment and mathematical formulation and they are not due to any physical condition. In fact, Fröhlich's theory was extended to include the cases of anisotropicity and nonlinearity [38]: this will be discussed in the chapter "Generalized Fröhlich Entropy: Anisotropic and Nonlinear Media".

3.2 *Formalism of the Theory*

All physical quantities are here assumed for unit of volume.[8] Textbooks on electromagnetism provide the expression for the electromagnetic energy U_e per unit volume of an ideal (loss-free) dielectric, is[9]

$$U_e = \frac{1}{2}\mathbf{E} \cdot \mathbf{D}. \tag{18}$$

If a variation in the temperature T causes a variation in d, in principle, from Eq. (18), we may extract the energy variation (per unit volume) of the considered dielectric medium. If the change of d is due to some thermodynamic transformation, the effective energy U_e will be a function of the details of the process. We note that, if we are in adiabatic conditions, the variation of this energy, dU_e, will coincide with the variation of the (total) internal thermodynamic energy dU [5, 7]. Nevertheless, adiabatic conditions are in most cases very difficult to be experimentally implemented [7]. Hence, Fröhlich's theory takes into account isothermal conditions, where dU_e coincides with the variation of the Helmholtz free energy dF [7, 11]. This is true because the volume variation of the system is here assumed to be negligible, so the variations of the Helmholtz and Gibbs free energies coincide. Thus, we treat a system with a negligible volume variation, where temperature T and the bias electric field $\mathbf{E}$ are the sole independent variables.

The linear dependence of the electric displacement vector D on the electric field E is expressed by the relationship

$$D = \varepsilon_0 \varepsilon E, \tag{19}$$

where ε_0 is the vacuum dielectric permittivity and ε is a scalar quantity (scalar approximation). These last assumptions will be overcome by the generalized form of the equations, which has been developed by extending Fröhlich's method [38] and will be discussed in the chapter "Generalized Fröhlich Entropy: Anisotropic and Nonlinear Media". However, in order to highlight the foundational elements, we here strictly follow Fröhlich's exposition. The system is considered both without losses, so ε is purely real, and without volume variation.

---

8 As in Fröhlich's treatise.

9 See Eq. 10 in Sect. 3.1 in chapter "Introduction: Dielectric Calorimetry and Fröhlich Entropy Estimation".

The scalar form of the standard relationship providing the density of electric energy u_e is

$$u_e = \frac{1}{2} E D. \tag{20}$$

When a static field is applied we have

$$u_e = \frac{1}{2} \varepsilon_0 \varepsilon_s E^2, \tag{21}$$

where ε_0 is the static vacuum dielectric constant and ε_s is the relative static dielectric function. When ε_s changes with temperature, Eq. (21) provides the energy variation (per unit volume) of the considered dielectric. From Eq. (20), we consider the quantity

$$du_e = \frac{1}{2} E dD, \tag{22}$$

that is the energy variation of the dielectric when D is infinitesimally varied. If some thermodynamic transformation produces a change in D, the amount of energy given by the Relationships (20) and (21) will depend on the character of the considered thermodynamic process. Thus, we have to take into account the two fundamental laws of thermodynamics, which were previously expressed in Relationships (5) and (1). For an ideal system with volume v and pressure p which experiences a reversible transformation, the first and second laws take the form

$$dU = dQ - pdv, \tag{23}$$

$$dS = \frac{dQ}{T}, \tag{24}$$

where Q is the heat exchanged by the system, U is the exchanged energy, S is the entropy (all calculated per unit volume). We here recall the definition of the Helmholtz free energy F as

$$F = U - TS, \tag{25}$$

that provides the maximum amount of work which can be obtained from a system in an isothermal process.[10] Equations (22) and (23) give the variation of energy dU during a process in which the field, or the temperature, or both, are slightly changed:

$$dU = dQ + \frac{1}{2} E dD. \tag{26}$$

[10] We stress that Fröhlich considers a system with a constant volume and just T and E can vary.

This expression has a mathematical structure akin to that of Eq. (23), if pressure and volume are respectively replaced by $-E$ and D. However, D in Eq. (26) depends on temperature, so, by applying the hypothesis of Eq. (19), we have

$$dD = d(\varepsilon_0 \varepsilon_s E) = \varepsilon_0 \left(\varepsilon_s dE + E \frac{\partial \varepsilon_s}{\partial T} dT \right),$$

whose meaning is that a variation of D is composed of two terms: a variation of field-strength E at constant temperature, and a variation of temperature at constant E. Here, it is useful to use T and E^2 as the independent variables, so we define

$$\mathcal{E} = E^2$$

and we accomplish the change of variable congruently. The first law of thermodynamics given by Eq. (26) will become:

$$\begin{aligned} dU &= dQ + \frac{1}{2} \varepsilon_0 \varepsilon_s d\mathcal{E} + \mathcal{E} \varepsilon_0 \frac{\partial \varepsilon_s}{\partial T} dT = \\ &= \frac{\partial U}{\partial \mathcal{E}} d\mathcal{E} + \frac{\partial U}{\partial T} dT. \end{aligned} \tag{27}$$

A second relation is obtained from the second principle, by imposing that dS, given by Eq. (24), is a total differential. This implies that there exists a unique function $S = S(T, \mathcal{E})$ such that

$$dS = \frac{\partial S}{\partial T} dT + \frac{\partial S}{\partial \mathcal{E}} d\mathcal{E}. \tag{28}$$

So, we have

$$dS = \mathcal{A}(T, \mathcal{E}) dT + \mathcal{B}(T, \mathcal{E}) d\mathcal{E}, \tag{29}$$

where $\mathcal{A}$ and $\mathcal{B}$ are both functions of T and $\mathcal{E}$. If we impose the condition that dS is a total differential we obtain

$$\frac{\partial \mathcal{B}}{\partial T} = \frac{\partial^2 S}{\partial T \partial \mathcal{E}} = \frac{\partial \mathcal{A}}{\partial \mathcal{E}}. \tag{30}$$

Now, after inserting dQ from (27) into relationships (23) and (24), we have

$$dS = \frac{1}{T} \left(\frac{\partial U}{\partial T} - \varepsilon_0 \mathcal{E} \frac{\partial \varepsilon_s}{\partial T} \right) dT + \frac{1}{T} \left(\frac{\partial U}{\partial \mathcal{E}} - \frac{\varepsilon_0 \varepsilon_s}{2} \mathcal{E} \right). \tag{31}$$

Such a kind of equation is the same as (29); then Eq. (30) will be

$$\frac{\partial}{\partial T}\left[\frac{1}{T}\left(\frac{\partial U}{\partial \mathcal{E}}-\frac{\varepsilon_0\varepsilon_s}{2}\mathcal{E}\right)\right]=\frac{\partial}{\partial \mathcal{E}}\left[\frac{1}{T}\left(\frac{\partial U}{\partial T}-\varepsilon_0\mathcal{E}\frac{\partial \varepsilon_s}{\partial T}\right)\right].$$

If we derive this last one we will obtain

$$\frac{\partial U}{\partial \mathcal{E}}=\frac{\varepsilon_0}{2}\left(\varepsilon_s+T\frac{\partial \varepsilon_s}{\partial T}\right),$$

whose integration

$$U=\int\frac{\partial U}{\partial \mathcal{E}}d\mathcal{E}=\int\frac{\varepsilon_0}{2}\left(\varepsilon_s+T\frac{\partial \varepsilon_s}{\partial T}\right)d\mathcal{E}.$$

will provide the total internal energy (density) up to an additive constant not depending on $\mathcal{E}=E^2$

$$U=U_0+\left(\varepsilon_s+T\frac{\partial \varepsilon_s}{\partial T}\right)\frac{\varepsilon_0 E^2}{2}. \tag{32}$$

Here $U_0=U_0(T)$, although independent of E^2, it depends on T, because it provides the energy of the dielectric at zero field.

Through these relationships, the entropy S may be easily calculated: comparing Eq. (31) with Eq. (28), both $\partial S/\partial T$ and $\partial S/\partial \mathcal{E}$ are known if U is introduced from Eq. (32). Thus

$$\frac{\partial S}{\partial T}=\frac{1}{T}\frac{\partial U_0}{\partial T}+\frac{\varepsilon_0\mathcal{E}}{2}\frac{\partial^2\varepsilon_s}{\partial T^2},$$
$$\frac{\partial S}{\partial \mathcal{E}}=\frac{\varepsilon_0}{2}\frac{\partial \varepsilon_s}{\partial T},$$

and, from

$$S=\int\frac{\partial S}{\partial \mathcal{E}}d\mathcal{E}=\int\frac{\varepsilon_0}{2}\frac{\partial \varepsilon_s}{\partial T}d\mathcal{E},$$

we may obtain

$$S=S_0+\frac{\varepsilon_0 E^2}{2}\frac{\partial \varepsilon_s}{\partial T}, \tag{33}$$

where $S_0 = S_0(T)$ is the entropy at zero field. By applying the definition of Eq. (25) we also may calculate a similar expression for the Helmholtz free energy

$$F = F_0 + \frac{\varepsilon_0 \varepsilon_s E^2}{2}, \tag{34}$$

where $F_0 = F_0(T)$ is the free energy in the absence of a field. Ultimately, the relationships derived from Fröhlich's treatise show that, given the assumptions introduced earlier, they take the following mathematical form:

$$U(T, E) = U_0(T) + \frac{1}{2}\left(\varepsilon + T\frac{\partial \varepsilon}{\partial T}\right) E^2, \tag{35a}$$

$$S(T, E) = S_0(T) + \frac{1}{2}\frac{\partial \varepsilon}{\partial T} E^2, \tag{35b}$$

$$F(T, E) = F_0(T) + \frac{1}{2}\varepsilon E^2, \tag{35c}$$

which are the *Fröhlich's thermodynamic relationships* [11], i.e. respectively *Fröhlich total energy*, *Fröhlich entropy*, *Fröhlich free energy* (always for unitary volume) [40–42]. We stress that the (35) pinpoint that the standard formula (21) does provide the Helmholtz free energy (F) and not the total internal energy U.

4 Phase Transitions

4.1 Overview

Among the principal applications of dielectric thermodynamics is the analysis and classification of phase transitions. Phase transitions constitute a central and extensively investigated topic in condensed matter physics [43–45]. Accordingly, we restrict our discussion to a concise and essential overview, strictly aligned with the thematic scope of this monograph. We focus on two classes, namely ferroelectric solids and polar and nematic liquids, of dielectric materials in which the study of phase transitions is pivotal for understanding their physical behavior, both from a fundamental standpoint and in view of their relevance for applications.

4.2 Ferroelectrics and Relaxors

Ferroelectrics are solid dielectric materials that exhibit spontaneous electric polarization below the Curie temperature, accompanied by a hysteresis loop and a corresponding mechanical strain. A feature shared by the ferroelectric effect and a related phenomenon—the pyroelectric effect—is their pronounced temperature dependence: both effects manifest only below a specific threshold temperature.

However, the ferroelectric effect differs from the pyroelectric one in that the former supports reversible polarization [14].

The passage between the ferroelectric and non-ferroelectric state is among the most extensively investigated phase transitions. Such a transition involves changes in several microscopic and mesoscopic features of the considered materials, often involving long-range symmetries and transformation of crystalline structures. However, it can be fruitfully studied and described from a macroscopic point of view. This is done by analyzing the variations of the dielectric response as a function of temperature or, more rarely, some other thermodynamic parameter. If the electric susceptibility of a given ferroelectric is plotted as a function of temperature, the transition is typically indicated by the presence of a peak [2]: below the peak temperature, the system resides in the *ferroelectric phase*; above it, in the *paraelectric phase*. The intermediate temperature region, where the transition effectively occurs, may be either broad or narrow, depending strongly on the intrinsic properties of the material. The classification of this regime—often involving the identification of one or more characteristic temperatures—is a topic widely addressed in the literature [46].

Ferroelectric materials are generally categorized into 2 principal classes, with several intermediate states: *ferroelectrics in the strict sense* and *relaxors*. The former exhibit a sharp $\varepsilon = \varepsilon(T)$ peak at the transition (Fig. 3a), and their response

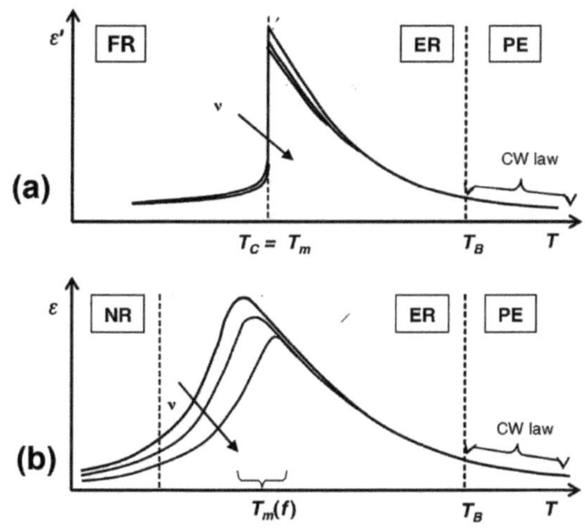

Fig. 3 Adapted from Fig. 9 of [47]. Ideal schemes of different possibilities for the temperature evolution of structure and dielectric properties of dielectrics. The plots show the behavior of the real part of the dielectric constant as a function of temperature, $\varepsilon = \varepsilon(T)$, with the frequency parameter ν. (**a**) canonical ferroelectric with a sharp phase transition; (**b**) canonical relaxor with a diffuse relaxor-to-ferroelectric phase. The Curie-Weiss temperature T_C, maximum of ε temperature T_m, Burns temperature T_B. Several regions are highlighted: standard ferroelectric (FR), normal relaxor (NR), transition range ("ergodic relaxor") (ER); finally, the paraelectric region (PE), where Curie-Weiss law holds. (Courtesy of the original publisher ©)

shows negligible dependence on the frequency ν. In contrast, relaxors display a significantly broader $\varepsilon = \varepsilon(T)$ curve around the transition (Fig. 3b), along with a non-negligible—often substantial—frequency dependence. At sufficiently high temperatures (namely above the so-called *Burns temperature* T_B) both classes cease to exhibit ferroelectric behavior and enter the *paraelectric phase* ("PE" regime in Fig. 3), where they obey the standard Curie–Weiss law[11] [2, 14, 47].

Thermodynamics is particularly useful to investigate and describe the macroscopic behavior of a material in proximity to some of its transition temperatures [12, 14]. Starting from this approach, the free energy landscape is usually analyzed in several works where the phenomenological investigation of the ferroelectric transition is considered. Actually, the specific form of free energy employed depends on the imposed external constraints; under typical conditions—namely, constant temperature, negligible mechanical stress, and absence of external electric fields—Gibbs free energy is particularly convenient, as it is conventionally formulated as a function of temperature T, mechanical stress σ, and electric polarization P.[12] Thus, if $\sigma = 0$, $G = G(T, P)$ will be

$$G(T, P) \simeq G_0(T) + \frac{1}{2}g_2 P^2 + \frac{1}{4}g_4 P^4 + \frac{1}{6}g_6 P^6, \quad (36)$$

where the expansion coefficients g_2, g_4, g_6 are temperature-dependent. From Rel. (36) we may calculate the corresponding electric field

$$E \simeq \frac{\partial G}{\partial P} = g_2 P + g_4 P^3 + g_6 P^5. \quad (37)$$

The character of the phase transition is determined by the sign of g_2, g_4, g_6 (or higher order coefficients), displaying patterns which can also be quite complicated.

To give a very simple example, it is worth considering here a very elementary scenario, already discussed by A. F. Devonshire to provide a general idea [12]. In this case, all coefficients except $g_2 = g_2(T)$ are positive, and P varies continuously across the transition temperature. Such condition coincides with a case of a second-order phase transition. The minima of $G(T, P)$ are calculated through $\frac{\partial G}{\partial P} = 0$; the corresponding pattern is plotted in Fig. 4. The different curves displayed in Fig. 4a correspond to different values of $g_2 = g_2(T)$ (assumed as a parameter). In the vicinity of the transition temperature, which usually coincides with the Curie temperature T_C [46], the coefficient g_2 can be modeled as a linear function of temperature [12]:

[11] See Eq. (14) in Sect. 3.3 in chapter "Introduction: Dielectric Calorimetry and Fröhlich Entropy Estimation".

[12] For the sake of simplicity, we here consider a scalar approximation of the vectorial and tensorial fields. This simplification suffices to describe phase transitions in large classes of ferroelectrics, as well as the upper transitions in perovskite systems. However, it does not capture the lower transitions in perovskites, which entail a reorientation of the polarization direction.

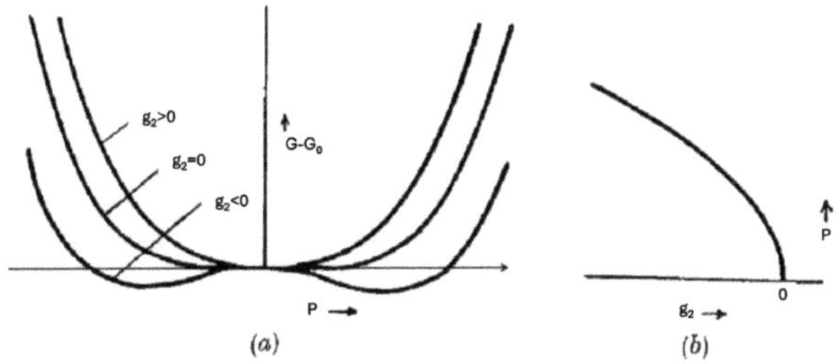

Fig. 4 Adapted from Fig. 5 of [12]. Second-order transition. (**a**) Free energy G as a function of polarization at the transition temperature and at temperatures just above and just below. (**b**) Spontaneous polarization as a function of temperature at a second-order transition. (Courtesy of the original publisher ©)

Fig. 5 Adapted from Fig. 6 of [12]. First-order transition. (**a**) Free energy G as a function of polarization at the transition temperature and at temperatures just above and just below. (**b**) Polarization as a function of temperature at a first-order transition. (Courtesy of the original publisher ©)

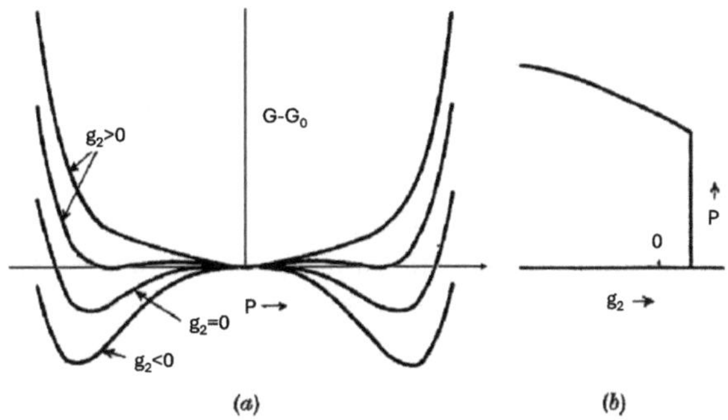

$$g_2 = \mu(T - T_C),$$

with $\mu > 0$. So, g_2 spans negative to positive values: the $g_2 = 0$ curve in Fig. 4a corresponds to the $T = T_C$ temperature. The plot shows that P evolves continuously from zero at T_C to a nonzero value as the temperature decreases below T_C.

These hypotheses require that the system undergoes a second-order phase transition, which is characterized by the absence of latent heat and a discontinuity in the specific heat capacity. Conversely, if any of the higher-order expansion coefficients assume negative values, the resulting free energy profile will exhibit a more complex topology, as shown in Fig. 5.

Here we note that the presented thermodynamic approach can also be used to explain many ferroelectric phenomena, related to both second-order and first-order transitions.

4.3 Polar and Mesogenic Liquids

Understanding how intermolecular interactions govern the structural organization of a liquid is one of the central challenges in the molecular physics of the liquid state.

Among the many open issues within this wide physical problem, the investigation of phase transitions is one of the most actively investigated topics. While the occurrence of a transition from a liquid to a glassy state is one of the greatest open problems [17, 18, 20, 48], another, correlated, specific issue is the transition between the isotropic (disordered) state, where molecules are essentially uncorrelated, and the organized (ordered) state with the arising of some symmetry breaking, where molecules are strongly interrelated. Two large classes of compounds, with several analogies between them, are usually considered, i.e. polar liquids and nematogenic liquids. Both of them display phase transitions, from the isotropic liquid to the nematic liquid crystal, as well as those occurring between distinct liquid crystalline phases, ranking among the most striking phenomena observed in physics [49]. The general pattern of transitions is a very broad topic. We here concentrate on the specific use of dielectric thermodynamics.

Indeed, the analysis of the dielectric response is one of the classical experimental techniques to analyze the occurrence of such manifold kinds of transitions [13, 15, 50–55]. The use of a thermodynamic theoretical analysis of the measured dielectric response was demonstrated to be a fruitful tool to inspect phase transitions in such classes of liquids. These compounds are paradigmatic of the fact that the peculiar isotropic-to-ordered transition is often experimentally triggered not by varying the temperature T, but by the applied electric field E. Therefore, both these independent variables are used to drive the physical response.

In the study of such liquids, the results concerning the temperature dependence of the (static) dielectric response ε_s and its derivative are important because they pinpoint the evolution of the basic thermodynamic quantities in the investigated system. A standard application of such an approach is the use of Fröhlich relationships (35c) and (35b) to pinpoint the phase transitions from an uncorrelated to a correlated molecular [13]. In nematogenic liquids, the I-N transition, from the isotropic (I) to the nematic (N) phase, can be easily highlighted. Following this approach, the increments of the electric field-induced Helmholtz free energy ΔF and the entropy ΔS can be calculated from Eq. (35c) and (35b); so, we have:

$$\frac{\Delta F}{E^2} = \frac{F(T,E) - F_0(T)}{E^2} = \frac{\varepsilon_0}{2}\varepsilon_s, \tag{38a}$$

$$\frac{\Delta S}{E^2} = \frac{S(T,E) - S_0(T)}{E^2} = \frac{\varepsilon_0}{2}\frac{\partial \varepsilon_s}{\partial T}. \tag{38b}$$

Therefore, the quantitative analysis of the Helmholtz free energy and Fröhlich entropy (FE) provides an effective framework for characterizing the isotropic-to–nematic (I–N) phase transition. A representative application of this approach is illustrated by the data in Fig. 6, adapted from [15], where the I–N transition is identified through the observed variations in Helmholtz free energy ΔF and Fröhlich entropy ΔS. Specifically, a negative shift in ΔS (i.e., $\Delta S < 0$) reflects the typical response of dipolar liquids, wherein the application of an external electric field E enhances molecular ordering within the system.

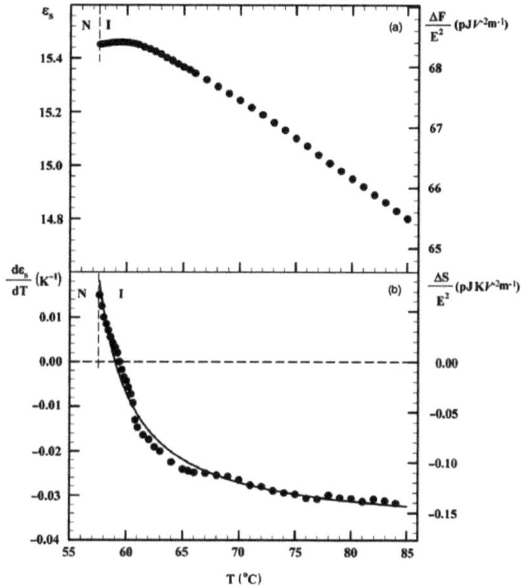

Fig. 6 From Fig. 2 of [15]. Temperature dependence of the static dielectric permittivity (**a**) and its derivative (**b**) (left axes) and the corresponding Helmholtz free energy ΔF and the entropy ΔS increments (right axes) induced by the probing electric field in the isotropic phase of nematogenic organic liquid 4-cyanophenyl-40-n-alkylbenzoate ($C_7H_{15}PhCOOPhCN$, 7CPB). (Courtesy of the original publisher ©)

For highly polar mesogenic compounds exhibiting a behavior like in Fig. 6, the entropy increment ΔS becomes positive prior to the onset of the I–N transition. This indicates that the electric field, when applied to the isotropic dipolar phase, disrupts molecular ordering at the microscopic level. Such a field-induced increase in ΔS is feasible only if the dipolar mesogenic molecules exhibit partial preorganization in the isotropic phase, predominantly in an antiparallel configuration. Under these conditions, the pre-existing intermolecular structures—namely, microscopic domains with nematic-like orientational order—are partially dismantled by the electric field, which promotes parallel dipole alignment.

Consequently, the system undergoes a competition between the intrinsic self-assembly of mesogenic molecules in the pre-nematic regime and the externally imposed orientational bias due to the electric field. As the temperature decreases and approaches the I-N transition threshold, this interplay intensifies, manifesting a behavior reminiscent of critical phenomena.

A similar behavior of Helmholtz free energy and Fröhlich entropy, as highlighted by Relations (38), can be found in isotropic-to-organized phase transitions occurring in several polar and nematogenic liquids, making such analysis a trustworthy approach in these investigations [13, 15, 39, 50–54].

5 Elements of Thermodynamics of Not Ideal Dielectrics

5.1 Configurational Entropy

The Helmholtz free energy F of a system at absolute temperature T is given by Eq. (25), i.e.

$$F = U - TS$$

We previously recalled that Fröhlich worked out the increase in entropy due to the application of an external electric field, so that the fundamental Relationship (33) holds [10, 11], i.e.

$$S = S_0(T) + \frac{\varepsilon_0}{2}\frac{\partial \varepsilon_s}{\partial T}E^2$$

We note that in the derivation of such formula, exposed in Sect. 3.2, it is inherently assumed that the imaginary part of the complex dielectric function $\varepsilon^* = \varepsilon' - i\varepsilon''$, i.e. the loss due to dielectric relaxation processes or conductivity contributions, can be neglected and only the real part ε' has to be considered; a conjecture which in soft matter systems usually is not fulfilled.

Johari [17, 18, 20, 21, 56, 57] discussed in a series of papers the impact of this "Fröhlich entropy" on glassy systems and differentiates additionally between configurational and vibrational contributions, and concludes that the latter are considered negligible; hence

$$\Delta S_{\mathrm{conf}} = \frac{\varepsilon_0}{2}\frac{d\varepsilon_s}{dT}E^2 \tag{39}$$

The quantity $\Delta S_{\mathrm{conf}}/E^2$ can have negative or positive values attributed to *type I* or *type II* liquids (see Table 4). For the former it is assumed that the application of

Table 4 The quantities $\partial\varepsilon_s/\partial T$ and the calculated $\Delta S^E_{\mathrm{conf}}/E^2$ for several liquids at a temperature far above their respective glass transition temperature T_g. For *type I* liquids, $\partial\varepsilon_s/\partial T$ and $\Delta S^E_{\mathrm{conf}}/E^2$ are negative while for *type II* liquids, these are positive. Data taken from [20]

Material	T_g [K]	$\partial\varepsilon_s/\partial T$ [K^{-1}]	$\Delta S_{\mathrm{conf}}/E^2$ [$J\ K^{-1}\ cm^{-1}\ V^{-2}$]
Acrylnitrile	223	−0.27	−1.195
N-methylacetonitrile	308	−0.14	−0.620
N-ethylacetonitrile	238	−3.2	−14.162
Propylene glycol	186	−0.3	−1.328
Glycerol	218	−0.6	−2.655
Ethanol	134	−0.73	−3.231
Water	259	−0.46	−2.036
2-Methyl-3-heptanol	273	0.011	0.049
3-Methyl-3-heptanol	288	0.012	0.053

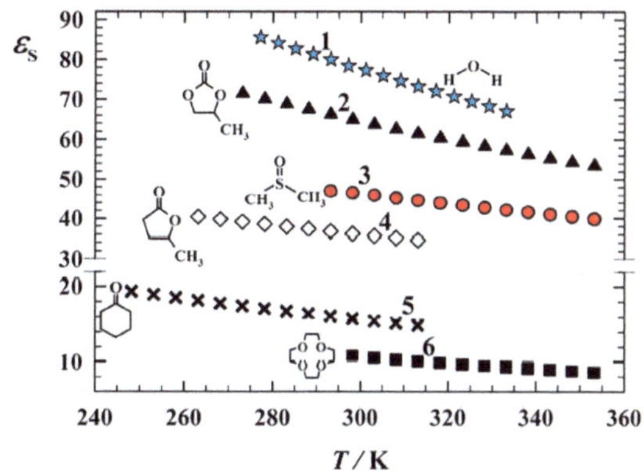

Fig. 7 Temperature dependences of the static permittivity of 1—water (the dipole moment of single molecule amounts 1.84 D), 2—propylene carbonate (4.94 D), 3—dimethylsulfoxide (3.96 D), 4—γ-valerolactone (4.24D), 5—cyclohexane (3.08 D), 6—12-crown-4 (2.33 D). Reproduced from [13] with permission. (Courtesy of the original publisher ©)

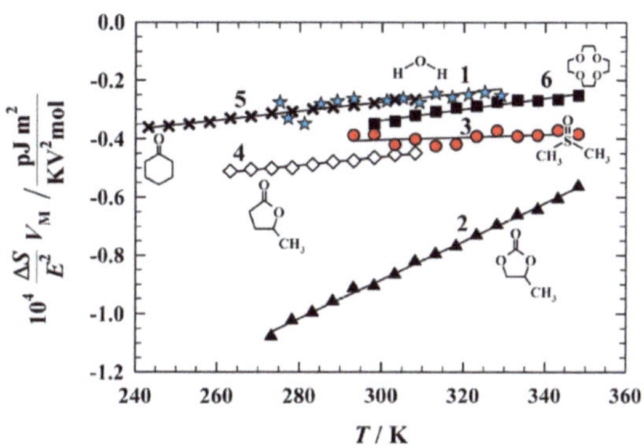

Fig. 8 Temperature dependences of the increment of the molar oriental entropy of 1—water (the dipole moment of single molecule amounts 1.84 D), 2—propylene carbonate (4.94 D), 3—dimethylsulfoxide (3.96 D), 4—γ-valerolactone (4.24D), 5—cyclohexane (3.08 D), 6—12-crown-4 (2.33 D). Reproduced from [13] with permission. (Courtesy of the original publisher ©)

the external electric field E increases configurational order while for the latter this seems not to be the case. A molecular interpretation is not given.

Świergiel et al. [13] measured for a variety of molecular liquids the temperature dependence of the static permittivity $\varepsilon_s(T)$ (Fig. 7) and of the increment of molecular entropy $\Delta\varepsilon_s/E^2$ (Fig. 8). The former is interpreted as the result of a parallel ordering of polar molecules leading to an increase of polarity in the liquid, the latter effect is rationalized similarly, because due to an increased polarity the impact of the external electric field is enhanced and hence the increment of entropy decreases. Hereby water and DMSO play a special role due to their character of H-bond formation and their molecular structure, respectively.

5.2 Nonlinear Contributions

In the past two decades, nonlinear Dielectric Spectroscopy has made enormous progress as summarized in [58]. Based on refined measurement techniques, it became possible to determine the third and fifth harmonic of the dielectric susceptibility

$$P = \frac{\mu_{\rm dip}}{a^d} \quad \mathcal{L}_d\left[\frac{\mu_{\rm dip}E_{\rm st}}{k_BT}\right] \simeq$$

$$\simeq \frac{1}{3}\frac{\mu_{\rm dip}}{a^d}\left(\frac{\mu_{\rm dip}E_{\rm st}}{k_BT}\right) - \frac{1}{45}\frac{\mu_{\rm dip}}{a^d}\left(\frac{\mu_{\rm dip}E_{\rm st}}{k_BT}\right)^3 + \frac{2}{945}\frac{\mu_{\rm dip}}{a^d}\left(\frac{\mu_{\rm dip}E_{\rm st}}{k_BT}\right)^5 + \ldots \tag{40}$$

where $E_{\rm st}$ is the static electric field applied to molecules with dipole moment $\mu_{\rm dip}$, a^d is the molecular d-dimensional volume, $\mathcal{L}_d$ is the corresponding Langevin function expressing the thermal equilibrium of a single dipole in interaction with the external electric field; the numerical pre-factors used in Eq. (40) of the first-, third-, and fifth- order responses correspond to the case $d = 3$ [59]. Glassy densification is characterized by a dramatic slowing down of the molecular dynamics. This is mostly interpreted as consequence of a growing length scale of amorphously ordered regions. Mathematically, it can be described by an approach developed by Bouchard and Brioli (BB) [60, 61] in which thermodynamic fluctuations correlate with collective rearrangement over a length-scale increasing with decreasing temperature. The explicit formulation is quite elaborate and results in specific features for the third and fifth harmonic of the dielectric susceptibility while the linear response remains unaffected (Fig. 9) [62, 63].

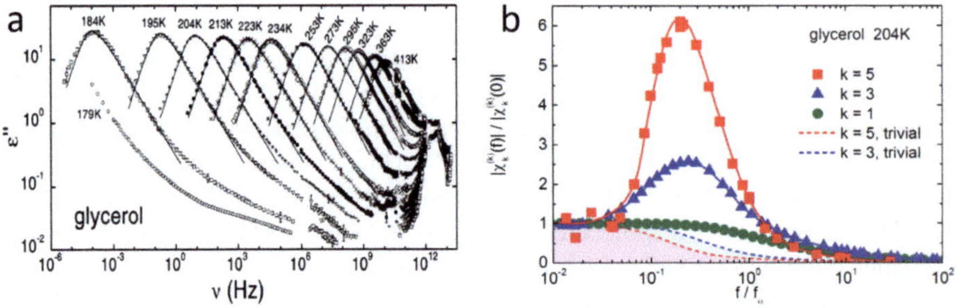

Fig. 9 Linear and nonlinear dielectric spectra in molecular liquids. (**a**) For glycerol, wide frequency range dielectric spectrum for the imaginary part of the linear dielectric susceptibility: the relaxation frequency f_a -where $\varepsilon'' \equiv \chi_1''$ is maximum- strongly varies in T. Note that, because χ_1' supersedes χ_1'' around f_a, the modulus $|\chi_1|$ of the linear response has no peak and continuously decreases when frequency increases—see the green points in part **b** of this figure. (**b**) Moduli of the linear ($k = 1$), third order ($k = 3$) and fifth order ($k = 5$) susceptibility in glycerol at 204 K which amounts to $T/T_g \simeq 1.08$ since glycerol $T_g \simeq 188$ K. The trivial case of an ideal gas of non-interacting molecules yields monotonically decreasing responses when increasing frequency, at all order k in the field. Reproduced from [63] with permission. (Courtesy of the original publisher ©)

In summary, we may say that an external electric field acting on a material causes multiple thermodynamic effects on entropy phase transitions, relaxation time and in the case of polar liquids the glass transition temperature. The analysis of higher harmonic in the non-linear dielectric response delivers furthermore information concerning the length scale on which glassy dynamics takes place.

References

1. S. Burns, J. Am. Ceram. Soc. **104**, 2087 (2021)
2. K. Kao, *Dielectric Phenomena in Solids - With Emphasis on Physical Concepts of Electronic Processes* (Elsevier Academic Press, San Diego, CA, 2004)
3. F. Kremer, A. Schönhals (eds.), *Broadband Dielectric Spectroscopy* (Springer, Berlin, Heidelberg, 2003)
4. F. Kremer, A. Loidl (eds.), *The Scaling of Relaxation Processes* (Springer, Berlin, Heidelberg, 2018)
5. M. Abraham, R. Becker, *The Classical Theory of Electricity and Magnetism* (Blackie & Son Limited, London, 1932)
6. A. Devonshire, Philos. Mag. **40**, 1040 (1949)
7. R. Becker, *Electromagnetic Fields and Interactions* (Blackie & Son, London, 1964)
8. R. Newhman, *Properties of Materials - Anisotropy, Symmetry, Structure* (Oxford University Press, Oxford, 2005)
9. J. Nye, *Physical Properties of Crystals - Their Representation by Tensors and Matrices* (Claredon Press, Oxford, 1985)
10. H. Fröhlich, *Theory of Dielectrics*, 1st edn. (Claredon Press, Oxford, 1949)
11. H. Fröhlich, *Theory of Dielectrics*, 2nd edn. (Claredon Press, Oxford, 1958)
12. A. Devonshire, Adv. Phys. **3**, 85 (1954)
13. J. Świergiel, I. Płowaś, J. Jadżyn, J. Mol. Liq. **220**, 879 (2016)
14. F. Jona, G. Shirane, *Ferroelectric Crystals* (Dover, New York, 1993)
15. J. Jadżyn, J. Swiergiel, D. Bauman, R. Dąbrowski, K. Czupryński, U. Sokołowska, Phase Transit. **82**, 409 (2009)
16. W. Kipnusu, E. Zhuravlev, C. Schick, F. Kremer, Macromol. Chem. Phys. **224**, 2200443 (2023)
17. G. Johari, J. Chem. Phys. **112**, 7518 (2000)
18. G. Johari, J. Chem. Phys. **112**, 8958 (2000)
19. S. Samanta, R. Richert, J. Chem. Phys. **142**, 044504 (2015)
20. G. Johari, J. Chem. Phys. **138**, 154503 (2013)
21. G. Johari, J. Chem. Phys. **145**, 164502 (2016)
22. H. Zhang, F. Wang, B. Nestler, J. Comput. Phys. **505**, 112907 (2024)
23. H. Zhang, F. Wang, B. Nestler, J. Chem. Phys. **161**, 044704 (2024)
24. D.V. Matyushov, J. Chem. Phys. **145**, 034504 (2016)
25. S. Samanta, R. Richert, J. Phys. Chem. B **120**, 7737 (2016)
26. S. de Groot, P. Mazur, *Non-equilibrium Thermodynamics*, reprint edition (Dover Publications, New York, 1984). Originally published by North-Holland in 1962
27. G. Maugin, *Non-equilibrium Thermodynamics of Electromagnetic Solids* (Springer Wien, Wien, 1993), pp. 205–258
28. H. Callen, *Thermodynamics and an Introduction to Thermostatistics* (John Wiley & Sons, New York, 1985)
29. L. Landau, E. Lifshitz, *Statistical Physics - Part 1*, 3rd edn. (Pergamon Press, Oxford, UK, 1980)
30. G. Heckmann, *Ergebnisse der exakten Naturwissenschaften - Springer Tracts in Modern Physics*, vol. 4 (Springer, Berlin), p. 100 (1925)

31. S. Baoyuan, W. Jiantong, Z. Jun, Q. Min, J. Mater. Process. Technol. **139**(1), 444 (2003)
32. J. Scott, Annu. Rev. Mater. Res. **41**, 229 (2011)
33. L. Landau, E. Lifshitz, *Electrodynamics of Continuous Media* (Pergamon Press, Oxford, UK, 1960)
34. V. Daniel, *Dielectric Relaxation* (Academic Press, London, 1967)
35. C. Böttcher, *Theory of Electric Polarization*, 2nd edn. (Elsevier Science, Amsterdam, 1973)
36. B. Scaife, *Principles of Dielectrics* (Claredon Press, Oxford, 1998)
37. L. Brillouin, *Science and Information Theory* (Academic Press, New York, 1962)
38. J. Parravicini, Phys. B **541**, 54 (2018)
39. J. Parravicini, G. Parravicini, Res. Phys. **28**, 104571 (2021)
40. J. Parravicini, E. DelRe, A. Agranat, G. Parravicini, Phys. Rev. B **93**, 094203 (2016)
41. J. Parravicini, E. DelRe, A. Agranat, G. Parravicini, Nanoscale **9**, 9572 (2017)
42. P. Tan, H. Tian, F. Huang, X. Meng, Y. Wang, C. Hu, X. Cao, L. Li, Z. Zhou, Phys. Rev. Appl. **11**, 024037 (2019)
43. M. Fisher, Rep. Prog. Phys. **30**, 615 (1967)
44. E. Stanley, *Introduction to Phase Transitions and Critical Phenomena* (Claredon Press - OUP, Oxford, UK, 1971)
45. G. Jaeger, Arch. Hist. Exact Sci. **53**, 51 (1998)
46. J. Toulouse, Ferroelectrics **369**(1), 203 (2008)
47. A. Bokov, Z. Ye, Front. Ferroelectr. **41**, 31 (2006)
48. L. Leuzzi, T. Nieuwenhuizen, *Thermodynamics of the Glassy State* (Taylor & Francis - CRC Press, Boca Raton, 2007)
49. S. Singh, Phys. Rep. **324**, 107 (2000)
50. J. Jadżyn, G. Czechowski, J. Phys. Chem. B **111**, 3727 (2007)
51. J. Jadżyn, G. Czechowski, J.L. Déjardin, M. Ginovska, J. Phys. Chem. A **111**, 8325 (2007)
52. J. Jadżyn, J.L. Déjardin, M. Ginovska, Acta Phys. Pol. A **111**, 877 (2007)
53. J. Jadżyn, U. Sokołowska, G. Czechowski, J. Phys. Chem. B **112**, 7022 (2008)
54. J. Jadżyn, G. Czechowski, J. Déjardin, J. Phys. Chem. B **112**, 4948 (2008)
55. T. Ezquerra, A. Nogales (eds.), *Crystallization as Studied by Broadband Dielectric Spectroscopy* (Springer, Berlin, Heidelberg, 2020)
56. G.P. Johari, G. Power, J.K. Vij, J. Chem. Phys. **117**, 1714 (2002)
57. G. Johari, J. Chem. Phys. **139**, 026101 (2013)
58. R. Richert (ed.), *Nonlinear Dieletric Spectroscopy* (Springer, Berlin, Heidelberg, 2018)
59. S. Albert, M. Michl, P. Lunkenheimer, A. Loidl, P. Déjardin, F. Ladieu, Third and fifth harmonic responses in viscous liquids, in *Nonlinear Dielectric Spectroscopy*, ed. by R. Richert (Springer, Cham, 2018), pp. 219–260
60. J.P. Bouchaud, G. Biroli, Phys. Rev. B **72**, 064204 (2005)
61. L. Berthier, G. Biroli, J. Bouchaud, L. Cipelletti, W. van Saarloos, *Dynamical Heterogeneities in Glasses, Colloids, and Granular Media* (Oxford University Press, Oxford, UK, 2011)
62. S. Albert, T. Bauer, M. Michl, G. Biroli, J.P. Bouchaud, A. Loidl, P. Lunkenheimer, R. Tourbot, C. Wiertel-Gasquet, F. Ladieu, Science **352**, 1308 (2016)
63. O. Dauchot, F. Ladieu, C. Royall, C. R. Phys. **24**, 25 (2023)

Generalized Fröhlich Entropy: Anisotropic and Nonlinear Media

Jacopo Parravicini

Abstract Starting from the Fröhlich's theory, the thermodynamic potentials are analytically given in generalized forms, comprehending the case of a dielectric material with either anisotropic or nonlinear response. The obtained relationships can be employed for describing different classes of anisotropic materials. Two peculiar cases are also considered: the possibility of highlighting an hyperbolic thermodynamic-dielectric response when specific anisotropic features are involved; the direct calculation of nonlinear coefficients for a physical system within a Debye-Langevin model.

1 Introduction

In condensed matter physics, the main mathematical theories describing important macroscopic phenomena such as electrocaloric, pyroelectric, piezocaloric, magnetocaloric, etc. effects, have been widely available for decades [1–3]. These correlate thermodynamic variables, which are typically scalar, with physical quantities that are typically vectorial and tensorial; in this context thermodynamic potentials and variables are expressed as a function of e.g. electric and magnetic fields, susceptibility tensors, dielectric displacement and magnetization fields, stresses and strains tensors, etc. [3–6]. This framework makes it possible, inter alia, to generalize the thermodynamic relations of Maxwell to the case of materials with dielectric, magnetic, and elastic anisotropies [2, 4]. From a broad perspective one may state that in condensed matter physics the vast majority of the macroscopic properties of a material admit a representation "by tensors and matrices"[1] [3, 4, 7–9].

[1] J. Nye in [3].

J. Parravicini (✉)
Dipartimento di Fisica & Astronomia, Università degli Studi di Firenze, Sesto F., Firenze, Italy
e-mail: jacopo.parravicini@unifi.it

J. Parravicini (ed.), *Fröhlich Entropy Estimation and Dielectric Calorimetry*,
Advances in Dielectrics, https://doi.org/10.1007/978-3-032-21454-6_3

When we focus on the thermodynamic of dielectrics, the previously discussed Fröhlich's theory may be regarded as a natural starting point [10–12]. Nevertheless, the original formulation of it [10, 11], which would be recalled by the subsequent scholars of dielectrics [7, 13–16], always took into account the simplest assumptions about the material under investigation. As already discussed in the previous chapter "Introduction: Dielectric Calorimetry and Fröhlich Entropy Estimation", while the considered thermodynamic conditions must be *reversible* and *isothermal* processes, the conditions about the electromagnetic response of the material are:

- isotropic response;
- linear dependence of **D** on **E**.

Although the development of a theory providing the thermodynamic fundamental equations of a dielectric with not-negligible nonlinearities and anisotropies has not been addressed, such a generalization is important in order to place Fröhlich's basic theory within the framework of the representation of physical systems by tensors and matrices. This is further motivated by the fact that Fröhlich's theory, especially *Fröhlich entropy evaluation*, has been applied to a wide variety of systems exhibiting (at least in principle) both anisotropies and nonlinearities, such as nematicons [17–22], inorganic ferroelectric and disordered perovskites [23–25], polar glasses [26], polar liquids [27–31], polymeric films [32], enzymes [33]. In short we may say that the validity of the isotropy and linearity assumptions, which Fröhlich's equations are based on, strongly depends on the specific dielectric features. Namely, when considering systems with an intrinsically directional response such as an isotropic crystals [24, 25, 34, 35] and nematicons [17–20], we will need to take into account anisotropy; moreover, in the investigation of dipolar liquids [27], the linearity assumed by Fröhlich strictly holds just when $E \rightarrow 0$, requiring to evaluate the amount of possible not negligible nonlinear corrections.

Here we intend to address these issues in order to develop a thermodynamic theory analogous to that by Fröhlich which is still valid for anisotropic and nonlinear materials. We start from the basic equations already discussed in chapter "Introduction: Dielectric Calorimetry and Fröhlich Entropy Estimation" (Eqs. (5) and (9) in chapter "Introduction: Dielectric Calorimetry and Fröhlich Entropy Estimation") to obtain the equivalent of expressions (9) when either anisotropy or nonlinearity, or both, are present. Similarly to the original Fröhlich's treatise, all physical quantities are here assumed to be defined per unit volume. Apart from isotropy and linearity, all other Fröhlich's basic hypotheses (previously listed in Sect. 3.1 in chapter "Introduction: Dielectric Calorimetry and Fröhlich Entropy Estimation") are assumed to remain valid.

2 Case of Anisotropic Linear Media

2.1 General Relationships

When an isotropic linear dielectric is considered, the real permittivity ε is a 2-rank tensor, i.e. $\boldsymbol{\varepsilon} \equiv \varepsilon_{ij}$. The correlation between the applied static electric field $\mathbf{E}$ and the displacement vector $\mathbf{D}$ has the form $\mathbf{D} = \boldsymbol{\varepsilon} \cdot \mathbf{E}$. If we label the components as $E_x = E_1$, $E_y = E_2$, $E_z = E_3$, and similarly for $\mathbf{D}$, we may write

$$D_i = \sum_j \varepsilon_{ij} E_j. \tag{1}$$

We consider the variation of U_e given by [11]:

$$dU_e = \mathbf{E} \cdot d\mathbf{D}, \tag{2}$$

which provides the variation of energy (per unit volume) when $\mathbf{D}$ infinitesimally varies. By combining Eq. (2) with the first principle of thermodynamics, we obtain the variation of energy dU for a thermodynamic process where the field or the temperature, or both, are slightly (reversibly) changed,:

$$dU = dQ + dU_e = dQ + \mathbf{E} \cdot d\mathbf{D}. \tag{3}$$

where dQ is the amount of exchanged heat (for unit of volume) during the transformation. The differential of $\mathbf{D}$ for an anisotropic dielectric is

$$dD_i = d\Big(\sum_j \varepsilon_{ij} dE_j\Big) = \sum_j \left(\varepsilon_{ij} dE_j + E_j \frac{\partial \varepsilon_{ij}}{\partial T} dT\right). \tag{4}$$

The internal energy U will take the form

$$U = U(T, \mathbf{E}) = U(T, E_1, E_2, E_3).$$

Placing Eq. (4) into Eq. (3) and applying the differential theorem, we get

$$dU = dQ + \sum_i E_i \left(\sum_j \varepsilon_{ij} dE_j + E_j \frac{d\varepsilon_{ij}}{dT} dT\right) = \sum_j \frac{\partial U}{\partial E_j} dE_j + \frac{\partial U}{\partial T} dT.$$

Likewise, applying to this last one the definition of entropy in reversible transformations, $dS = dQ/T$, the entropy S can be written as a differential relationship, i.e.

$$dS = \frac{1}{T}\left[\frac{\partial U}{\partial T} - \sum_i E_i \left(\sum_j E_j \frac{\partial \varepsilon_{ij}}{\partial T}\right)\right] dT + \frac{1}{T}\sum_i \left(\frac{\partial U}{\partial E_i} - \sum_j E_j \varepsilon_{ij}\right) dE_i. \tag{5}$$

The reversibility condition requires that dU and dS are exact differentials, then we can apply the Schwarz theorem. Therefore, we have

$$\frac{\partial^2 S}{\partial E_j \partial E_i} = \frac{\partial^2 S}{\partial E_i \partial E_j} \quad \text{and} \quad \frac{\partial^2 S}{\partial T \partial E_i} = \frac{\partial^2 S}{\partial E_i \partial T} \tag{6}$$

for E_i and $i, j = 1, 2, 3$, and analogously for E_j and $j, i = 1, 2, 3$. If we combine relationships (6) with Eq. (5), we will have

$$\frac{\partial}{\partial T}\left[\frac{1}{T}\sum_i \left(\frac{\partial U}{\partial E_i} - \sum_j E_j \varepsilon_{ij}\right)\right] = \frac{\partial}{\partial E_i}\left\{\frac{1}{T}\left[\frac{\partial U}{\partial T} - \sum_i E_i \left(\sum_j E_j \frac{\partial \varepsilon_{ij}}{\partial T}\right)\right]\right\},$$

which gives

$$\frac{\partial U}{\partial E_i} = \sum_j E_j \left(\varepsilon_{ij} + T\frac{\partial \varepsilon_{ij}}{\partial T}\right),$$

whose integration

$$U = \int \frac{\partial U}{\partial E_i} dE_i = \int \sum_j E_j \left(\varepsilon_{ij} + T\frac{\partial \varepsilon_{ij}}{\partial T}\right) dE_i$$

gives the energy density on the involved thermodynamic quantities:

$$U = U_0(T) + \frac{1}{2}\sum_{ij} E_j \left(\varepsilon_{ij} + T\frac{\partial \varepsilon_{ij}}{\partial T}\right) E_i. \tag{7}$$

Here $U_0(T)$ is an integration constant independent of E_i and it represents the total temperature-dependent energy of the dielectric at zero electric field.

In the same way we may obtain similar expressions for entropy and Helmholtz free energy: by inserting Eq. (7) into Eq. (5) and taking into account the differential theorem, we obtain

$$\frac{\partial S}{\partial T} = \frac{1}{T}\frac{\partial U_0}{\partial T} + \frac{1}{2}\sum_{ij} E_j \frac{\partial^2 \varepsilon_{ij}}{\partial T^2} E_i$$

$$\text{and for } i = 1, 2, 3 \tag{8}$$

$$\frac{\partial S}{\partial E_i} = \sum_j E_j \frac{\partial \varepsilon_{ij}}{\partial T}.$$

So, from

$$S = \int \frac{\partial S}{\partial E_i} dE_i = \int \sum_j E_j \frac{\partial \varepsilon_{ij}}{\partial T} dE_i$$

we obtain

$$S = S_0(T) + \frac{1}{2} \sum_{ij} \left(E_j \frac{\partial \varepsilon_{ij}}{\partial T} \right) E_i, \tag{9}$$

where $S_0(T)$ is the part of the entropy which does not depend on **E**. Finally, from relationship (7) and (9) we calculate the free energy $F = U - TS$, i.e.

$$F = F_0(T) + \frac{1}{2} \sum_{ij} \left(\varepsilon_{ij} E_j \right) E_i, \tag{10}$$

where, of course, $F_0(T)$ is the free energy at 0 field.

Relationships (7), (9), and (10) are the thermodynamic relationships for a linear anisotropic dielectric analogous to the Fröhlich's ones for an isotropic material [12, 36].

2.2 Remarks and Discussion

The relations obtained here provide the connections between the considered thermodynamic potentials and the dielectric physical quantities, where the tensorial form of the dielectric permittivity highlights the anisotropy. We note that in the isotropic limit, when ε is a scalar quantity, (7), (9), and (10) coincide with the original relationships (9) provided by Fröhlich. We also note that the static conditions which we are considering make not possible to have different frequencies to distinguish different fields, so field-mixing phenomena are not allowed, contrary to the typical phenomena of nonlinear optics [37]. Hence, the different indices which label the electric field components E_i and E_j are just a mathematical notation to handle the not-diagonal $\boldsymbol{\varepsilon}$ 2-rank tensor, without a real physical meaning, so in effective calculation one should assume $E_i = E_j$ for $i, j = 1, 2, 3$.

2.3 Tensorial Representation

Expressions (7), (9), and (10) can be viewed from a more general perspective by taking into account that physical properties of materials, namely crystals, are conventionally and most naturally represented by means of tensors and matrices. First, we may define the following quantities:

$$u_{ij} \equiv \frac{1}{2}\left(\varepsilon_{ij} + T\frac{\partial \varepsilon_{ij}}{\partial T}\right), \tag{11a}$$

$$s_{ij} \equiv \frac{1}{2}\left(\frac{\partial \varepsilon_{ij}}{\partial T}\right), \tag{11b}$$

$$f_{ij} \equiv \frac{\varepsilon_{ij}}{2}, \tag{11c}$$

which are relationships defining three tensors, whose components are the field-induced parts of respectively total energy, entropy, and free energy for unitary field at a given temperature T. By exploiting Eqs. (11), (7), (9), and (10) can be expressed in the following compact form:

$$U = U_0(T) + \sum_{ij} E_j u_{ij} E_i, \tag{12a}$$

$$S = S_0(T) + \sum_{ij} E_j s_{ij} E_i, \tag{12b}$$

$$F = F_0(T) + \sum_{ij} E_j f_{ij} E_i., \tag{12c}$$

i.e.[2]

$$U_E = \sum_{ij} E_j u_{ij} E_i, \tag{13a}$$

$$S_E = \sum_{ij} E_j s_{ij} E_i, \tag{13b}$$

$$F_E = \sum_{ij} E_j f_{ij} E_i. \tag{13c}$$

This form of the relations highlights that under anisotropic conditions U_E, S_E, F_E may be computed, at a given T, by means of the coefficients u_{ij}, s_{ij}, f_{ij} respectively.

[2] Taking into account Eqs. (5) of chapter "Introduction: Dielectric Calorimetry and Fröhlich Entropy Estimation".

Here tensorial physical quantities are associated to the thermodynamic potentials: a closely related formalism is encountered in the theories of stress and strain, as well as in pyroelectric, piezoelectric, and piezocaloric effects [3, 4].

The advantage of introducing the coefficients u_{ij}, s_{ij}, f_{ij} in Eqs. (12) and (13) is that, when the electric field is applied along a given direction, the corresponding variations of thermodynamic potentials U_E, S_E, F_E can be computed through a contraction of the associated tensor with the square of the applied field. This point can be emphasized by assuming a reference frame aligned with the principal axes. Accordingly, the generic dielectric tensor

$$\varepsilon_{ij} = \begin{pmatrix} \varepsilon_{11} & \varepsilon_{12} & \varepsilon_{13} \\ \varepsilon_{21} & \varepsilon_{22} & \varepsilon_{23} \\ \varepsilon_{31} & \varepsilon_{32} & \varepsilon_{33} \end{pmatrix}, \tag{14}$$

in the principal reference frame, say x, y, z, has the diagonal form [3]

$$\varepsilon_{ij} = \begin{pmatrix} \varepsilon_{11} & 0 & 0 \\ 0 & \varepsilon_{22} & 0 \\ 0 & 0 & \varepsilon_{33} \end{pmatrix}. \tag{15}$$

This notation can be made even more compact and transparent by defining the components of the tensor along each principal direction, i.e.

$$\varepsilon_x \equiv \varepsilon_{11}, \qquad \varepsilon_y \equiv \varepsilon_{22}, \qquad \varepsilon_z \equiv \varepsilon_{33}, \tag{16}$$

so that, of course, $\varepsilon_{ij} \equiv \boldsymbol{\varepsilon}$ of (15) will be written in the form

$$\boldsymbol{\varepsilon} = \begin{pmatrix} \varepsilon_x & 0 & 0 \\ 0 & \varepsilon_y & 0 \\ 0 & 0 & \varepsilon_z \end{pmatrix}. \tag{17}$$

This implies that, in such a reference, the u_{ij}, s_{ij}, f_{ij} tensors[3] are also diagonal as tensor (15); the components of the first ones represent the change of the corresponding thermodynamic quantities for the direction where a unitary electric field is applied, i.e., at a given fixed temperature T, we have,

$$U_E(\mathbf{E}) = u_{11} E_x^2 + u_{22} E_y^2 + u_{33} E_z^2, \tag{18a}$$

$$S_E(\mathbf{E}) = s_{11} E_x^2 + s_{22} E_y^2 + s_{33} E_z^2, \tag{18b}$$

$$F_E(\mathbf{E}) = f_{11} E_x^2 + f_{22} E_y^2 + f_{33} E_z^2. \tag{18c}$$

[3] Defined by Eqs. (11).

The meaning of Eqs. (18) can be more easily appreciated when they are expressed in polar coordinates [36]. If **E** is written in terms of its strength E and the angular variables θ and ϕ, one obtains

$$U_E(E,\theta,\phi) = E^2\big(u_{11}\sin^2\theta\cos^2\phi + u_{22}\sin^2\theta\sin^2\phi + u_{33}\cos^2\theta\big), \tag{19a}$$

$$S_E(E,\theta,\phi) = E^2\big(s_{11}\sin^2\theta\cos^2\phi + s_{22}\sin^2\theta\sin^2\phi + s_{33}\cos^2\theta\big), \tag{19b}$$

$$F_E(E,\theta,\phi) = E^2\big(f_{11}\sin^2\theta\cos^2\phi + f_{22}\sin^2\theta\sin^2\phi + f_{33}\cos^2\theta\big). \tag{19c}$$

These relations make explicit the dependence on the field direction. They show that, when ε_{ij} is diagonal (Eq. (15)), the components of the tensors u_{ij}, s_{ij}, f_{ij} ((20a), (20b), (20c)) are the coefficients required to calculate variations of the considered thermodynamic potentials induced by the electric field of magnitude E and direction θ, ϕ. [4] This is analogous to what is found e.g. in the theory of piezoelectric and piezocaloric effects [3]. So that, by combining (15) with (18) or (19), we may make explicit a tensorial representation of coefficients defined by Relationships (11) in the principal axes reference frame at a given T, i.e.

$$u_{ij} = \begin{pmatrix} u_{11} & 0 & 0 \\ 0 & u_{22} & 0 \\ 0 & 0 & u_{33} \end{pmatrix} = \frac{1}{2}\begin{pmatrix} \varepsilon_{11} + T\frac{\partial\varepsilon_{11}}{\partial T} & 0 & 0 \\ 0 & \varepsilon_{22} + T\frac{\partial\varepsilon_{22}}{\partial T} & 0 \\ 0 & 0 & \varepsilon_{33} + T\frac{\partial\varepsilon_{33}}{\partial T} \end{pmatrix}, \tag{20a}$$

$$s_{ij} = \begin{pmatrix} s_{11} & 0 & 0 \\ 0 & s_{22} & 0 \\ 0 & 0 & s_{33} \end{pmatrix} = \frac{1}{2}\begin{pmatrix} \frac{\partial\varepsilon_{11}}{\partial T} & 0 & 0 \\ 0 & \frac{\partial\varepsilon_{22}}{\partial T} & 0 \\ 0 & 0 & \frac{\partial\varepsilon_{33}}{\partial T} \end{pmatrix}, \tag{20b}$$

$$f_{ij} = \begin{pmatrix} f_{11} & 0 & 0 \\ 0 & f_{22} & 0 \\ 0 & 0 & f_{33} \end{pmatrix} = \frac{1}{2}\begin{pmatrix} \varepsilon_{11} & 0 & 0 \\ 0 & \varepsilon_{22} & 0 \\ 0 & 0 & \varepsilon_{33} \end{pmatrix}. \tag{20c}$$

A practical example of the application of (20b), i.e. Eq. (11b) expressed in the principal-axes reference frame, is shown in Figs. 1 and 2 for a lithium-enriched potassium tantalate-niobate perovskite (KLTN). The considered sample is a z-cut single crystal, whose dielectric response $\varepsilon(T)$ is measured along the 3 principal axes (Fig. 1) [24]; the corresponding Fröhlich entropy (for a unit electric field) is displayed in Fig. 2, where it is calculated by exploiting the relations introduced above.

An equivalent interpretation of the coefficients u_{ij}, s_{ij}, f_{ij} is provided by Eqs. (19) which yield the resulting total energy, entropy, free energy, respectively, when a unit field along one of the principal directions of ε_{ij} is applied [3, 4, 24, 25]. These relations also indicate a practical way to evaluate the components of u_{ij}, s_{ij},

[4] The temperature dependence of u_{ij}, s_{ij}, f_{ij} was not specified to simplify the mathematical notation.

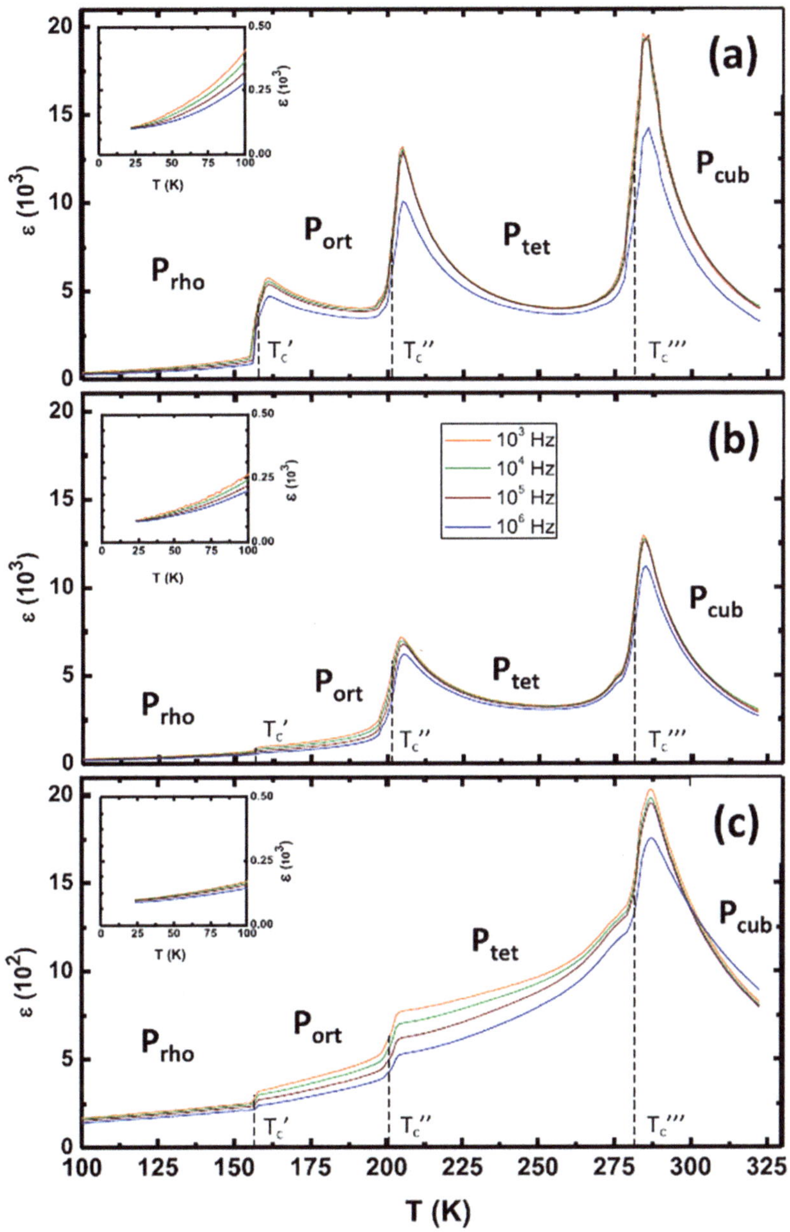

Fig. 1 Dependencies on temperature of the relative dielectric permittivity of KLTN single crystal measured at 10^3–10^6 Hz frequencies along a (**a**), b (**b**), and c (**c**) directions for the cooling stage. For each plot are indicated the nominal crystalline phases P_{rho} (rhombohedral), P_{ort} (orthorhombic), P_{tet} (tetragonal), and P_{cub} (cubic), with the relative transition temperatures T_C' (rhombohedral-to-orthorhombic), T_C'' (orthorhombic-to-tetragonal), and T_C''' (tetragonal-to-cubic). For each transition, the characteristic temperatures are found to be the same for all directions. Note that the insets show that the differences between permittivities tend to vanish as the temperature approaches 0 K. Reprinted with permission from [24]. (Courtesy of the original publisher © 2016, APS)

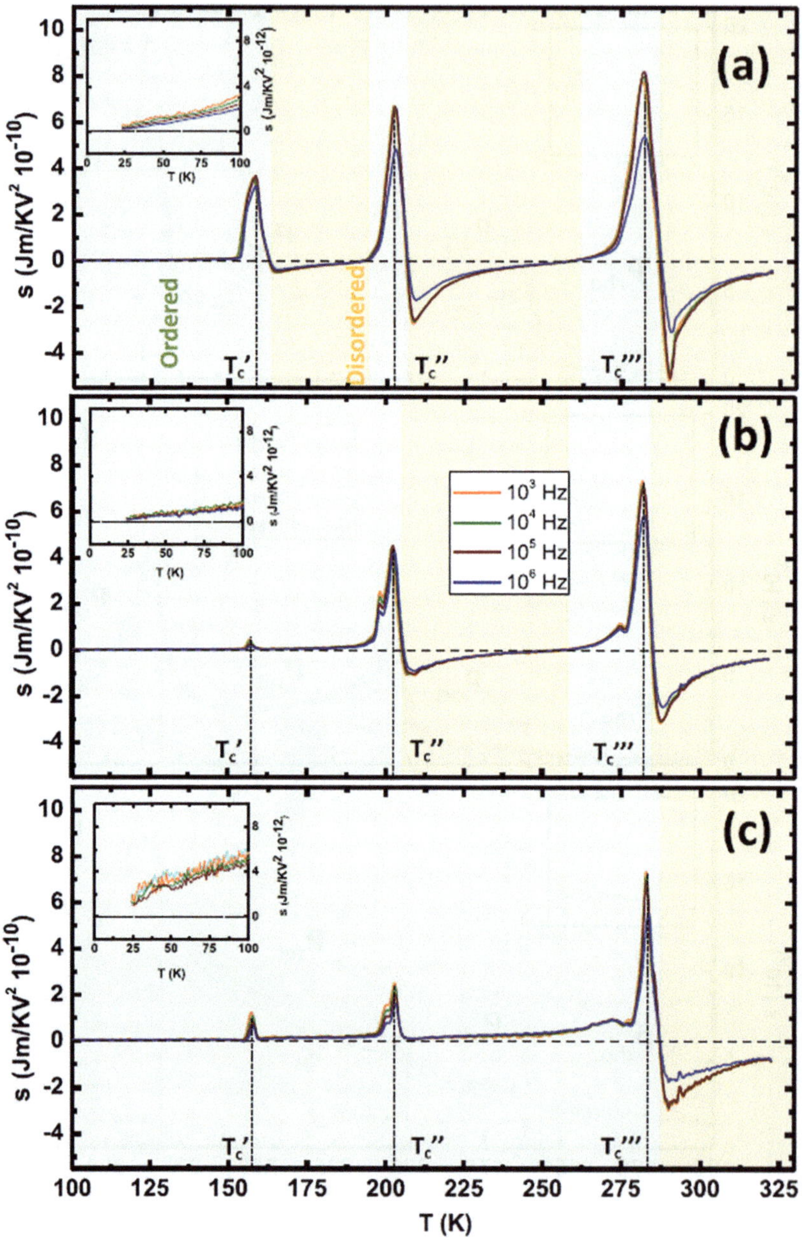

Fig. 2 Frölich entropy calculated from the 10 kHz data in Fig. 1 through Eq. (11b), indicated as s_a, s_b, s_c for, respectively, a (**a**), b (**b**), c (**c**) axes direction, for unit of volume and field square (cooling stage) in the 100–320 K temperature range. The indicated temperatures T_C', T_C'', T_C''' are the transition temperatures shown in Fig. 1. The insets show the temperature range 0–100 K for each respective direction and, congruently with the permittivity behavior, they point out that the anisotropy tend to vanish as the temperature approaches 0 K. Reprinted with permission from [24]. (Courtesy of the original publisher © 2016, APS)

f_{ij} tensors: they can be obtained by measuring ε and $\partial\varepsilon/\partial T$ along the main axis of the principal reference frame: after knowing the components of u_{ij}, s_{ij}, f_{ij} along the principal directions, the (19) or (18) can be used to deduce Fröhlich's total energy, entropy, free energy for an electric field **E** applied along an arbitrary direction [2, 3].

2.4 Geometrical Interpretation of Fröhlich's Entropy Coefficients

Let's now apply this formalism to the specific case of entropy. From (12), and specifically (20b), if we assume a reference frame in which dielectric tensor is diagonal, at a given temperature T, one finds

$$S_E(T, E) = s_{11}(T)E_x^2 + s_{22}(T)E_y^2 + s_{33}(T)E_z^2 \tag{21}$$

if we consider the usual quantity $s \equiv \frac{S_E}{E^2}$, that is Fröhlich entropy for unit of electric field, the above equation becomes

$$\frac{S_E}{E^2} = s = s_{11}\frac{E_x^2}{E^2} + s_{22}\frac{E_y^2}{E^2} + s_{33}\frac{E_z^2}{E^2} \tag{22}$$

By dividing each component, along the x, y, z directions, by the modulus E, it is useful to define the components, e_x, e_y, e_z, of the versor of the electric field **E**, i.e.

$$e_x \equiv \frac{E_x}{E}, \quad e_y \equiv \frac{E_y}{E}, \quad e_z \equiv \frac{E_z}{E}$$

so that Eq. (22) may be written in a compact form

$$s = s_{11}e_x^2 + s_{22}e_y^2 + s_{33}e_z^2, \tag{23}$$

and, defining the quantities $\sigma_x = 1/\sqrt{s_{11}}$, $\sigma_y = 1/\sqrt{s_{22}}$, $\sigma_y = 1/\sqrt{s_{33}}$, we have a quadric equation:

$$s = \frac{e_x^2}{\sigma_x^2} + \frac{e_y^2}{\sigma_y^2} + \frac{e_z^2}{\sigma_z^2}. \tag{24}$$

If the coefficients s_{ij} are all positive at the temperature T under consideration, Eq. (24) identifies an ellipsoid with semi-axes of lengths σ_x, σ_y, σ_z. Thus, the length of the tensor s_{ij} (20b) along the direction e_x is given by $\sigma_x = 1/\sqrt{s_{11}}$ and analogously for the directions e_y and e_z (Fig. 3). If, on the other hand, s_{11}, s_{22}, s_{33} are not all positive, coefficients σ_x, σ_y, σ_z are not all real, so that Eq. (24) identifies either a one-sheet or two-sheets hyperboloid [3] (Fig. 4). We also remember that, if

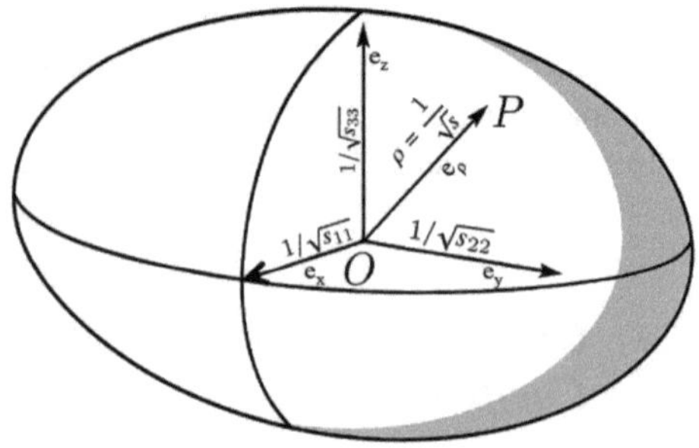

Fig. 3 Schematic of the ellipsoid given by Eq. (24) when all s_{ij} coefficients are positive

s_{ij} are all negative, (24) will be the equation of an imaginary ellipsoid. In all cases, if we consider a generic direction given by the versor e_ρ, we may define the quantities

$$\rho = \frac{1}{\sqrt{s}}, \quad s = \frac{1}{\rho^2},$$

thus Eq. (24) becomes

$$\frac{1}{\rho^2} = \frac{e_x^2}{\sigma_x^2} + \frac{e_y^2}{\sigma_y^2} + \frac{e_z^2}{\sigma_z^2}, \tag{25}$$

that is, given a generic radius vector ρ, with direction e_ρ, the magnitude s of the tensor s_{ij} in the quadric representation along that direction is equal to the reciprocal of the square length of ρ [3]. Therefore, the quadric surface given by Eq. (24) provides the geometrical representation of the tensor s_{ij}. Given a direction e_ρ, the length of the segment joining the origin to the intersection point P with this quadric surface is the reciprocal of the square root of the numerical value of the physical quantity s when the electric field is applied along e_ρ (Fig. 3).

We note that by multiplying Eq. (23) (or (24)) by E^2, one obtains Froehlich entropy S_E. This shows that Eq. (23) is equivalent, in fact, to the Relationship (19b) divided by E^2: for both (23) and (19b) give the value of Fröhlich entropy S_E, for a unitary field, in a given direction, at a fixed T. The advantage of writing (24) is the possibility of making a geometrical interpretation of the anisotropic response of the dielectric during a thermodynamic process. Indeed, the value of FE for unitary field, will be represented in the 3-dimensional space by a surface satisfying Eq. (24). Due to the basic Fröhlich's assumption of dealing isothermal processes,[5] the points of each surface all lie at the same temperature. We therefore conclude that Eq. (23),

$$s(T) = s_{11}(T)e_x^2 + s_{22}(T)e_y^2 + s_{33}(T)e_z^2,$$

[5] See Table 3, Sect. 3.1 in chapter "Elements of Thermodynamics of Dielectrics and Basic Theory of Fröhlich Entropy".

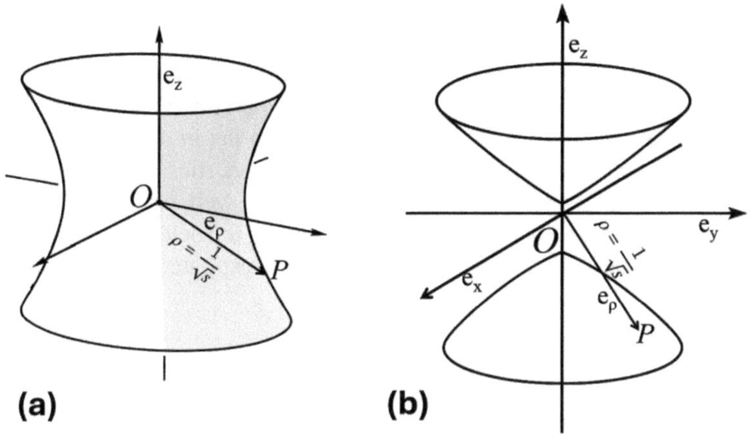

Fig. 4 Schematic of the one-sheet hyperboloid (**a**) and two-sheets hyperboloid (**b**) given by Eq. (24) when, respectively, one or two of the s_{ij} coefficients are negative. Sample e_ρ and ρ are also depicted

identifies a class of *isothermal surfaces* which, at a given fixed T, can display different geometrical and topological features.

In order to clarify such interpretation, it is worth applying the formalism to real data. We consider lithium-enriched potassium tantalate-niobate perovskite single-crystal (KLTN), whose dielectric response and Fröhlich entropy are respectively reported in Figs. 1 and 2 from [24]. This material system exhibits 4 nominal crystalline phases with marked anisotropies; in particular FE has qualitatively very different behaviors [24]. The first long-range phase is nominally rhombohedral (P_{rho} in Fig. 1), with positive values of FE along all the directions (Fig. 2). When we apply the geometrical interpretation of (24), the isothermal surface we obtain would be an ellipsoid similar to that in Fig. 3.

When we analyze the region $T > T_C'$, in the nominally orthorhombic phase (P_{ort} in Fig. 1), this considered system shows peculiar features. FE interpretation highlights the simultaneous presence of order and disorder at the same (fixed) temperature, as a function of the considered direction (Fig. 2). The evaluation FE as a function of the direction of applied field provides one negative (disordered-like configuration) and two positive (order-like configuration) values. Correlated coefficients of (23) will be $s_a < 0$, $s_b > 0$, and $s_c > 0$, so that, in our geometrical interpretation, Eq. (24) will identify a one-sheet hyperboloid surface (Fig. 4a). Therein, the openness of the obtained hyperboloid points out the non-existence of real solutions of Eq. (25) along some directions, i.e. there are some e_ρ for which such equation can not be satisfied for $\rho^2 > 0$.

The subsequent nominally tetragonal phase (P_{tet} in Fig. 1), in the temperature region $T > T_C''$, exhibits a further transition from an ordered-like to a disordered-like direction. This is evidenced by the presence of 2 negative coefficients in Eq. (23), so that along 2 principal axes no real solutions of Eq. (25) exists. In this condition Eqs. (24) identifies a two-sheets hyperboloid (Fig. 4b).

In summary, the geometrical interpretation allows us to highlight some not trivial features in anisotropic dielectrics. In the specific example we above discussed, such approach allows to highlight that phase transitions in disordered perovskites can determine a change of particular topological properties of the system, which are global and not local [24]. This phenomenon is well established, because it is confirmed also in another disordered anisotropic ferroelectric perovskite [25], so that it should deserve to be recognized and investigated in other anisotropic dielectrics.

3 Case of Nonlinear Media

3.1 General Relationships

We now consider the case where the dielectric displays a nonlinear response with respect to the applied electric field. This is the case, e.g., when some saturation effect is present, such as in Debye-Langevin physical systems (see Sect. 3.3 below). Under these conditions Eqs. (9), (7), (9), and (10) should be generalized. Without loos of generality, we here consider the case of second and third order nonlinearities: potential higher order terms can be obtained using the same theoretical framework and procedure. We start by writing the field-dependent terms of Fröhlich's relations $U = U(T, \mathbf{E})$, $S = S(T, \mathbf{E})$, $F = F(T, \mathbf{E})$ (as recalled in Eqs. (5)) in the form

$$U_E = U_E^{(1)} + U_E^{(2)} + U_E^{(3)}, \tag{26a}$$

$$S_E = S_E^{(1)} + S_E^{(2)} + S_E^{(3)}, \tag{26b}$$

$$F_E = F_E^{(1)} + F_E^{(2)} + F_E^{(3)}, \tag{26c}$$

where $U_E^{(1)}$, $U_E^{(2)}$, $U_E^{(3)}$ are, respectively, the term depending on linear, quadratic, and cubic susceptibilities, and analogously for entropy and free energy.

The fundamental relationship between the dielectric displacement $\mathbf{D}$ and the electric field $\mathbf{E}$ with nonlinear 2nd and 3rd order terms is

$$D_i = \sum_j \varepsilon_{ij}^{(1)} E_j + \sum_{jk} \varepsilon_{ijk}^{(2)} E_j E_k + \sum_{jkl} \varepsilon_{ijkl}^{(3)} E_j E_k E_l, \tag{27}$$

where $\varepsilon_{ij}^{(1)} = \varepsilon_{ij}$ of (1), $\varepsilon_{ijk}^{(2)}$ and $\varepsilon_{ijkl}^{(3)}$, for $i, j, k, l = 1, 2, 3$, are respectively the absolute second- and third- order nonlinear dielectric constants. The differential of D_i is

$$dD_i = \sum_j \left(\varepsilon_{ij}^{(1)} dE_j + \frac{\partial \varepsilon_{ij}^{(1)}}{\partial T} E_j dT \right) +$$
$$+ \sum_{jk} \left[\varepsilon_{ijk}^{(2)} \left(E_k dE_j + E_j dE_k \right) + \frac{\partial \varepsilon_{ijk}^{(2)}}{\partial T} E_j E_k dT \right] +$$
$$+ \sum_{jkl} \left[\varepsilon_{ijkl}^{(3)} \left(E_k E_l dE_j + E_j E_l dE_k + E_j E_k dE_l \right) + \frac{\partial \varepsilon_{ijkl}^{(3)}}{\partial T} E_j E_k E_l dT \right]. \tag{28}$$

So, the infinitesimal variation of U_e is

$$dU_e = \sum_{ij} \left[\varepsilon_{ij}^{(1)} \left(E_i dE_j + E_j dE_i \right) + E_i \frac{\partial \varepsilon_{ij}^{(1)}}{\partial T} E_j dT \right] +$$
$$+ \sum_{ijk} \left[\varepsilon_{ijk}^{(2)} \left(E_i E_k dE_j + E_i E_j dE_k + E_j E_k dE_i \right) + E_i \frac{\partial \varepsilon_{ijk}^{(2)}}{\partial T} E_j E_k dT \right] +$$
$$+ \sum_{ijkl} \left[\varepsilon_{ijkl}^{(3)} \left(E_i E_k E_l dE_j + E_j E_l dE_k + E_j E_k dE_l + E_j E_k E_l dE_i \right) + \right.$$
$$\left. + E_i \frac{\partial \varepsilon_{ijkl}^{(3)}}{\partial T} E_j E_k E_l dT \right], \tag{29}$$

which in scalar approximation would be

$$dU_e = \left(\varepsilon^{(1)} E + \varepsilon^{(2)} E^2 + \varepsilon^{(3)} E^3 \right) dE +$$
$$+ \left(\frac{\partial \varepsilon^{(1)}}{\partial T} E + \frac{\partial \varepsilon^{(2)}}{\partial T} E^2 + \frac{\partial \varepsilon^{(3)}}{\partial T} E^3 \right) dT.$$

By combining the (29) with (3) we can calculate the differential of U, i.e.

$$dU = dQ + dU_e =$$
$$= \sum_i \frac{\partial U}{\partial E_i} dE_i + \sum_j \frac{\partial U}{\partial E_j} dE_j + \sum_k \frac{\partial U}{\partial E_k} dE_k + \sum_l \frac{\partial U}{\partial E_l} dE_l + \frac{\partial U}{\partial T} dT, \tag{30}$$

and the corresponding entropy variation

$$dS = \frac{1}{T}(dU - dU_e) =$$

$$= \sum_i \frac{\partial S}{\partial E_i} dE_i + \sum_j \frac{\partial S}{\partial E_j} dE_j + \sum_k \frac{\partial S}{\partial E_k} dE_k + \sum_l \frac{\partial S}{\partial E_l} dE_l + \frac{\partial S}{\partial T} dT, \quad (31)$$

where dU_e is given by (29). The structure of Eq. (30) is analogous to Eq. (2.1) of the linear case, so we apply the differential theorem in the same way.

$$dS = \left\{ \frac{\partial U}{\partial T} - \sum_{ij} \left[\frac{\partial \varepsilon_{ij}^{(1)}}{\partial T} + \sum_k \left(\frac{\partial \varepsilon_{ijk}^{(2)}}{\partial T} + \sum_l \frac{\partial \varepsilon_{ijkl}^{(3)}}{\partial T} E_l \right) E_k \right] E_i E_j \right\} \frac{dT}{T} +$$
$$+ \sum_j \left\{ \frac{\partial U}{\partial E_j} - \sum_i \left[\varepsilon_{ij}^{(1)} + \sum_k \left(\varepsilon_{ijk}^{(2)} + \sum_l \varepsilon_{ijkl}^{(3)} E_l \right) E_k \right] E_i \right\} \frac{dE_j}{T}$$

In scalar approximation

$$dS = \frac{1}{T} \left[\frac{\partial U}{\partial T} - \left(\frac{\partial \varepsilon^{(1)}}{\partial T} + \frac{\partial \varepsilon^{(2)}}{\partial T} E + \frac{\partial \varepsilon^{(3)}}{\partial T} E^2 \right) E^2 \right] dT +$$
$$+ \frac{1}{T} \left[\frac{\partial U}{\partial E} - \left(\varepsilon^{(1)} + 2\varepsilon^{(2)} E + 3\varepsilon^{(3)} E^2 \right) E \right] dE$$

On the other hand, Eq. (31) is the nonlinear equivalent of Eq. (5); then, in the same hypothesis of a reversible transformation, we may apply the Schwarz theorem, so obtaining

$$\frac{\partial U}{\partial E_j} = \sum_i \left(\varepsilon_{ij}^{(1)} + T \frac{\partial \varepsilon_{ij}^{(1)}}{\partial T} \right) E_i + \sum_{ik} \left(\varepsilon_{ijk}^{(2)} + T \frac{\partial \varepsilon_{ijk}^{(2)}}{\partial T} \right) E_i E_k +$$
$$+ \sum_{ikl} \left(\varepsilon_{ijkl}^{(3)} + T \frac{\partial \varepsilon_{ijkl}^{(3)}}{\partial T} \right) E_i E_k E_l \quad (32)$$

or analogous expressions for $\partial U / \partial E_i$, $\partial U / \partial E_k$, $\partial U / \partial E_l$. The integral of Eq. (32)

$$U = \int \frac{\partial U}{\partial E_j} dE_j = \int \sum_i \left(\varepsilon_{ij}^{(1)} + T \frac{\partial \varepsilon_{ij}^{(1)}}{\partial T} \right) E_i dE_j +$$
$$+ \int \sum_{ik} \left(\varepsilon_{ijk}^{(2)} + T \frac{\partial \varepsilon_{ijk}^{(2)}}{\partial T} \right) E_i E_k dE_j + \int \sum_{ikl} \left(\varepsilon_{ijkl}^{(3)} + T \frac{\partial \varepsilon_{ijkl}^{(3)}}{\partial T} \right) E_i E_k E_l dE_j$$

provides the expression of the total energy for a nonlinear dielectric, i.e.

$$U = U_0(T) + \frac{1}{2}\sum_{ij}\left(\varepsilon_{ij}^{(1)} + T\frac{\partial \varepsilon_{ij}^{(1)}}{\partial T}\right) E_i E_j + \frac{1}{3}\sum_{ijk}\left(\varepsilon_{ijk}^{(2)} + T\frac{\partial \varepsilon_{ijk}^{(2)}}{\partial T}\right) E_i E_j E_k +$$
$$+\frac{1}{4}\sum_{ijkl}\left(\varepsilon_{ijkl}^{(3)} + T\frac{\partial \varepsilon_{ijkl}^{(3)}}{\partial T}\right) E_i E_j E_k E_l. \tag{33}$$

Therefore, after the same passages of (8), we integrate with respect to E_i similarly to (9) and (10), so obtaining the entropy

$$S = S_0(T) + \frac{1}{2}\sum_{ij}\left(\frac{\partial \varepsilon_{ij}^{(1)}}{\partial T}\right) E_i E_j +$$
$$+\frac{1}{3}\sum_{ijk}\left(\frac{\partial \varepsilon_{ijk}^{(2)}}{\partial T}\right) E_i E_j E_k + \frac{1}{4}\sum_{ijkl}\left(\frac{\partial \varepsilon_{ijkl}^{(3)}}{\partial T}\right) E_i E_j E_k E_l, \tag{34}$$

and, remembering that $F = U - ST$, from (33) and (34) we have the Helmholtz free energy

$$F = F_0(T) + \frac{1}{2}\sum_{ij}\left(\varepsilon_{ij}^{(1)}\right) E_i E_j +$$
$$+\frac{1}{3}\sum_{ijk}\left(\varepsilon_{ijk}^{(2)}\right) E_i E_j E_k + \frac{1}{4}\sum_{ijkl}\left(\varepsilon_{ijkl}^{(3)}\right) E_i E_j E_k E_l, \tag{35}$$

Relations (33), (34), and (35), retain the structure of the starting relations of Fröhlich's theory (Eqs. (5) in chapter "Introduction: Dielectric Calorimetry and Fröhlich Entropy Estimation") and their terms which depend on the field with the form of (26). These relationships are the most general formulation of the Fröhlich thermodynamic relationships, as they take into account the nonlinearities (of second and third order) as well as anisotropies [12, 36].

3.2 Remarks and Discussion

It is worth stressing the very general character of Eqs. (33), (34), and (35): provided that the above-recalled Fröhlich's assumptions are satisfied, they allow to calculate the thermodynamic response of all physical systems whose dielectric behavior can be described through Eq. (27). Moreover, they can also be employed when the second and third order terms of dielectric permittivity describe some field-induced anisotropy, i.e. where $\varepsilon^{(1)}$ is a scalar and $\varepsilon^{(2)}$ and $\varepsilon^{(3)}$ are tensors. Conversely, when the dielectric response cannot be cast in the form of (27), e.g. in the presence of non-monodromic behavior with respect to the electric field **E** (hysteresis effects), the relationships (33), (34), and (35) do not necessarily hold.

The linear terms of (33), (34), and (35) are associated with the linear susceptibility. They depend on the square of the electric field and, when the nonlinearity is negligible, they reduce to (7), (9), and (10), as expected. In our nonlinearity conditions, the assumption of a static regime (not dependence on time, 0 field frequency) makes no possible the arising of field-mixing phenomena such as in nonlinear optics [37], analogously to what was previously pinpointed for anisotropy conditions. Hence, the use of indices i, j, k, l in the equations are intended solely for computational purposes and not at indicating real physical effects.

The physical meaning of the just obtained expression (33), (34), and (35), can be underlined by writing them for isotropic conditions, i.e. when such relationships are given in scalar approximation. They are respectively:

$$U = U_0(T) + \frac{1}{2}\left(\varepsilon^{(1)} + T\frac{\partial \varepsilon^{(1)}}{\partial T}\right) E^2 +$$

$$+\frac{1}{3}\left(\varepsilon^{(2)} + T\frac{\partial \varepsilon^{(2)}}{\partial T}\right) E^3 + \frac{1}{4}\left(\varepsilon^{(3)} + T\frac{\partial \varepsilon^{(3)}}{\partial T}\right) E^4, \tag{36a}$$

$$S = S_0(T) + \frac{1}{2}\frac{\partial \varepsilon^{(1)}}{\partial T}E^2 + \frac{1}{3}\frac{\partial \varepsilon^{(2)}}{\partial T}E^3 + \frac{1}{4}\frac{\partial \varepsilon^{(3)}}{\partial T}E^4, \tag{36b}$$

$$F = F_0(T) + \frac{1}{2}\varepsilon^{(1)}E^2 + \frac{1}{3}\varepsilon^{(2)}E^3 + \frac{1}{4}\varepsilon^{(3)}E^4. \tag{36c}$$

These expressions highlight the presence of a second-order nonlinearity, which gives rise to a cubic dependence of the thermodynamic quantities on E. When an isothermal transformation is considered, this cubic dependence implies that a sign variation of E modifies the thermodynamic potentials.

This analysis shows that a non-zero second-order nonlinearity can not be found in a truly isotropic physical system: the system, instead, must be noncentrosymmetric. Hence, Eqs. (36) indicate that a material with a not-null static second-order nonlinearity must host a symmetry breaking. This is in remarkable agreement with the optical *second-order nonlinearity symmetry condition*: it requires that a not-null second-order susceptibility can be found only in a noncentrosymmetric material and it is widely known in nonlinear optics because it involves the second order nonlinear susceptibility at optical frequencies [37, 38]. Such a condition is usually demonstrated either by taking into account the features of a medium exhibiting a nonlinear response to an oscillating field [37] or by symmetry assumptions about the optical second-order susceptibility tensor [38]. Here relationships (33), (34), (35), and (36) give an independent demonstration of the same principle and confirm this second-order nonlinearity symmetry condition, whose they can be considered the extension for static (not optical) fields.

We finally note that our proposed approach can be straightforwardly extended to nonlinearity corrections of higher order by adding further power terms to Eq. (27)

and carrying out related calculations. This can be required to study physical systems like structural liquids, which can be described by models analogous to the Debye-Langevin one [27, 29, 39, 40], whose expansion contains the 3rd, 5th, and higher odd-order terms of the field. The expanded expressions will be polynomial extensions of (33), (34), (35) or (36). A specific example of such an approach is given in the following section.

3.3 *The Case of a Debye-Langevin System*

In the case of an isotropic physical system whose dielectric response can be described by the Debye-Langevin model, the dielectric response expansion coefficients can be easily analytically calculated. We start from the polarization P, for unit volume, as given by the Debye-Langevin function $\mathcal{L} = \mathcal{L}(a)$ [41]:

$$P = N\mu\mathcal{L}(a), \tag{37}$$

where N is the number of dipoles in a volume unit, μ is the elemental dipole moment, and [27, 39–41]

$$\mathcal{L}(a) = \coth(a) - \frac{1}{a}, \tag{38}$$

with $a = \frac{\mu E}{K_B T}$ (in the form of [41]). The expansion of this function is

$$\mathcal{L}(a) = \frac{1}{3}a - \frac{1}{45}a^3 + \frac{2}{945}a^5 - \frac{1}{4725}a^7 + \dots, \tag{39}$$

which has odd-order terms of $\mathbf{E}$ [27, 29]. Here we consider the terms until the seventh order. If we assume the constitutive relation of the polarization in scalar form, i.e.

$$P = D - \varepsilon_0 E$$

we may write a trivial expansion for a nonlinear dielectric

$$D = \varepsilon^{(1)}E + \varepsilon^{(2)}E^2 + \varepsilon^{(3)}E^3 + \varepsilon^{(4)}E^4 + \varepsilon^{(5)}E^5 + \varepsilon^{(6)}E^6 + \varepsilon^{(7)}E^7 + \dots. \tag{40}$$

By combining this with the (38) and limiting to the 7th order (which is wider than that in (27)), we obtain

$$P = N\mu\left[\frac{1}{3}\frac{\mu}{K_B T}E - \frac{1}{45}\frac{\mu^3}{(K_B T)^3}E^3 + \frac{2}{945}\frac{\mu^5}{(K_B T)^5}E^5 - \frac{1}{4725}\frac{\mu^7}{(K_B T)^7}E^7\right]$$

i.e.

$$P = \frac{N\mu^2}{3K_B T}E - \frac{N\mu^4}{45(K_B T)^3}E^3 + \frac{2N\mu^6}{945(K_B T)^5}E^5 - \frac{N\mu^8}{4725(K_B T)^7}E^7. \tag{41}$$

Congruently, if we express P as a function of D, following the (40), we will have

$$P = (\varepsilon^{(1)} - \varepsilon_0)E + \varepsilon^{(2)}E^2 + \varepsilon^{(3)}E^3 + \varepsilon^{(4)}E^4 + \varepsilon^{(5)}E^5 + \varepsilon^{(6)}E^6 + \varepsilon^{(7)}E^7. \tag{42}$$

Therefore, equalizing (41) with (42), we obtain

$$\begin{aligned} \varepsilon^{(1)} &= \frac{N\mu^2}{3K_B T} - \varepsilon_0, \qquad & \varepsilon^{(3)} &= -\frac{N\mu^4}{45(K_B T)^3} \\ \varepsilon^{(5)} &= \frac{2N\mu^6}{945(K_B T)^5}, \qquad & \varepsilon^{(7)} &= -\frac{N\mu^8}{4725(K_B T)^7}. \end{aligned} \tag{43}$$

Of course the factors of even powers of the filed will be all zero. So, by adding further expansion terms to the previously obtained relationships (36), we obtain the generalized scalar equations for the odd indices of ε

$$U = U_0(T) + \frac{1}{2}\left(\varepsilon^{(1)} + T\frac{\partial \varepsilon^{(1)}}{\partial T}\right)E^2 + \frac{1}{4}\left(\varepsilon^{(3)} + T\frac{\partial \varepsilon^{(3)}}{\partial T}\right)E^4 + \\ + \frac{1}{6}\left(\varepsilon^{(5)} + T\frac{\partial \varepsilon^{(5)}}{\partial T}\right)E^6 + \frac{1}{8}\left(\varepsilon^{(7)} + T\frac{\partial \varepsilon^{(7)}}{\partial T}\right)E^8 \tag{44a}$$

$$S = S_0(T) + \frac{1}{2}\frac{\partial \varepsilon^{(1)}}{\partial T}E^2 + \frac{1}{4}\frac{\partial \varepsilon^{(3)}}{\partial T}E^4 + \frac{1}{6}\frac{\partial \varepsilon^{(5)}}{\partial T}E^6 + \frac{1}{8}\frac{\partial \varepsilon^{(7)}}{\partial T}E^8 \tag{44b}$$

$$F = F_0(T) + \frac{1}{2}\varepsilon^{(1)}E^2 + \frac{1}{4}\varepsilon^{(3)}E^4 + \frac{1}{6}\varepsilon^{(5)}E^6 + \frac{1}{8}\varepsilon^{(7)}E^8. \tag{44c}$$

Finally, inserting (43) into (44), we have

$$U = U_0(T) - \frac{\varepsilon_0}{2}E^2 + \frac{N\mu^4}{90(K_B T)^3}E^4 - \frac{4N\mu^6}{2835(K_B T)^5}E^6 + \\ + \frac{N\mu^8}{6300(K_B T)^7}E^8 \tag{45a}$$

$$S = S_0(T) - \frac{N\mu^2}{6K_B T^2}E^2 + \frac{N\mu^4}{60K_B^3 T^4}E^4 - \frac{N\mu^6}{567K_B^5 T^6}E^6 + \\ + \frac{N\mu^8}{5400K_B^7 T^8}E^8 \tag{45b}$$

$$F = F_0(T) + \frac{1}{2}\left(\frac{N\mu^2}{3K_BT} - \varepsilon_0\right)E^2 - \frac{N\mu^4}{180(K_BT)^3}E^4 + \\ + \frac{N\mu^6}{2835(K_BT)^5}E^6 - \frac{N\mu^8}{37800(K_BT)^7}E^8. \quad (45c)$$

Therefore, once μ and N are known, these last practical relationships immediately provide the analytical form of the coefficients which can be directly inserted into Eqs. (36), conveniently expanded to higher orders, to perform direct calculations of U, S, F following the Debye-Langevin model.

4 Current and Possible Experimental Uses

4.1 Examples of Anisotropic Applications

Without claiming to be exhaustive of the physical systems, we briefly recall here some possible experimental applications of the here discussed equations. The first field of use of the anisotropic theory can be considered the description of crystalline solids, where specific tensors and matrices were largely developed to describe macroscopic properties [3, 5]. Past studies which exploited anisotropic Fröhlich entropy deal with anisotropic perovskites [24, 25, 35, 42]. The estimation of Fröhlich entropy is here accomplished to investigate the state-of-order around phase transitions, where the application of Eqs. (7), (9), (10), and (12) provides the mathematical tool to evaluate the entropic state. This allows us to follow the state-of-order along each generic crystalline direction as a function of the response to the applied field. Another possible approach, that has not already explored until now explored so far, is to exploit Eqs. (19) to *continuously* describe the energies and the state-of-order as a function of both direction and temperature. As perovskite-based materials have recently attracted considerable interest for energy-storage applications [43], a possible pathway could be, in principle, the realization and investigation of thermodynamic cycles as a function of direction of the applied electric field.

Another class of anisotropic in which where the Fröhlich's theory has been fruitfully employed is nematic and nematogenic liquids, where Fröhlich's relationships for the estimation of entropy and free energy were applied to highlight transition from isotropic to nematic phase [21]. In these systems there is a sharp distinction between the dielectric permittivity measured along the nematic axis, $\varepsilon_\parallel$, and the dielectric permittivity measured perpendicularly to it, $\varepsilon_\perp$: the figure of merit of the resulting anisotropy is the physical quantity $\Delta\varepsilon \equiv \varepsilon_\parallel - \varepsilon_\perp$. Several studies which exploited Fröhlich theory in these compounds decided to neglect the field-induced components of entropy and energy correlated with the dielectric anisotropy [17–21, 44]: the here exposed generalization, instead, provides the tools to evaluate the thermodynamic contributions associated to $\Delta\varepsilon$.

4.2 Examples of Nonlinear Applications

If we perform a dielectric thermal analysis of a sample, the experimental determination of $\varepsilon_{ij}^{(1)}$, $\varepsilon_{ijk}^{(2)}$, $\varepsilon_{ijkl}^{(3)}$ at different temperatures and, in the most general case, along different directions is the most straightforward method to practically obtain, through relationships (33), (34), and (35), the corresponding physical quantities. For example, a useful approach to estimate nonlinear terms is the measurement of the permittivity deviation from the linearity at different temperatures. When anisotropic dielectric media or materials with a nonlinear dielectric response are investigated, the theory above exposed, which is based on the generalization of Fröhlich's equations (Eqs. (9) in chapter "Introduction: Dielectric Calorimetry and Fröhlich Entropy Estimation"), allows us to have relationships which provide the corresponding thermodynamic physical quantities.

The necessity of a the generalization of Fröhlich's equations is highlighted by some works on structural liquids and glasses, which indicate the linearity of the dielectric response as a bottleneck to apply thermodynamic theory to such a kind of dielectrics [27–29, 45, 46]. The practical limits for the application of Fröhlich's theory is studied in detail in the work by G.P. Johari, where the field-induced entropy is analyzed in the experimental data of more than 40 structural liquids [27]. The final landscape highlights that the greatest part of the considered liquid systems can be described through basic linear Fröhlich's equations (Eqs. (9) in chapter "Introduction: Dielectric Calorimetry and Fröhlich Entropy Estimation") just around room temperature until a field of about 500 kV/cm, where the amount of nonlinear corrections are maintained within 1%. This means that either for higher values of field or at lower temperatures, the nonlinearities can not be neglected, hence the original basic Fröhlich's relationships (Eqs. (9)) cease to be valid. The nonlinear formulæ here discussed (33), (34), (35), and (36), on the other hand, take into account nonlinear corrections. The use of them removes the requirement of a linear dielectric response. Moreover, larger values of the electric field and corresponding lower values of temperature can be studied within the Fröhlich's framework, opening the possibility of using this formalism to explore phenomena correlated with high value of the field/temperature ratio.

5 Conclusions

When anisotropic or nonlinear dielectric materials are considered, Fröhlich's theory may be also used to derive thermodynamic potentials, provided that equations are suitably extended with terms taking into account anisotropic and nonlinear corrections. So the obtained relationships attain their most general form. In the limit of isotropic linear dielectrics, the generalized expressions become coincident to the original ones proposed by Fröhlich. The mathematical generalized formulæinclude physical quantities that can be experimentally measured, making them suitable for practical uses to obtain information from describe experimental results. The above discussed real physical systems are example of real possible experimental applications of such relationships. They demonstrate that this formalism is suitable to used to study large classes of dielectrics.

One advantage of such an approach lies in the relaxations of the basic assumptions to employ Fröhlich's theory: the obtained expressions allow us to characterize thermodynamic isothermal processes without stringent assumptions about the material. The generalization to anisotropic media point out the specific relations between thermodynamic variables and electric physical quantities, in conditions where the vectorial and tensorial nature of the electric responses are not negligible. Moreover, the generalization to nonlinear conditions, beyond the extension of the second-order nonlinearity symmetry condition to the zero-frequency fields (Eqs. (36) and Sect. 3.2) gives the possibility of calculating nonlinear corrections to the thermodynamic variables in dielectrics.

References

1. A. Devonshire, Philos. Mag. **40**, 1040 (1949)
2. A. Devonshire, Adv. Phys. **3**, 85 (1954)
3. J. Nye, *Physical Properties of Crystals - Their Representation by Tensors and Matrices* (Clarendon Press, Oxford, 1985)
4. R. Newhman, *Properties of Materials - Anisotropy, Symmetry, Structure* (Oxford University Press, Oxford, 2005)
5. J.R. Ray, J. Appl. Phys. **53**, 6441 (1982)
6. O. Narayan, A.P. Young, Am. J. Phys. **73**, 293 (2005)
7. L. Landau, E. Lifshitz, *Electrodynamics of Continuous Media* (Pergamon Press, Oxford, 1960)
8. P. Tsao, Am. J. Phys. **61**, 823 (1993)
9. H.J. Juretschke, Am. J. Phys. **45**, 277 (1977)
10. H. Fröhlich, *Theory of Dielectrics*, 1st edn. (Claredon Press, Oxford, 1949)
11. H. Fröhlich, *Theory of Dielectrics*, 2nd edn. (Claredon Press, Oxford, 1958)
12. J. Parravicini, G. Parravicini, Res. Phys. **28**, 104571 (2021)
13. R. Becker, *Electromagnetic Fields and Interactions* (Blackie and Son, London, 1964)
14. V. Daniel, *Dielectric Relaxation* (Academic Press, London, 1967)
15. C. Böttcher, *Theory of Electric Polarization*, 2nd edn. (Elsevier Science, Amsterdam, 1973)
16. B. Scaife, *Principles of Dielectrics* (Claredon Press, Oxford, 1998)
17. J. Jadżyn, G. Czechowski, J. Phys. Chem. B **111**, 3727 (2007)
18. J. Jadżyn, U. Sokołowska, J. Déjardin, J. Phys. Chem. B **112**, 9050 (2008)
19. J. Jadżyn, G. Czechowski, J. Déjardin, J. Phys. Chem. B **112**, 4948 (2008)
20. J. Jadżyn, J. Swiergiel, D. Bauman, R. Dąbrowski, K. Czupryński, U. Sokołowska, Phase Transit. **82**, 409 (2009)
21. D. Bauman, J. Jadżyn, E. Wolarz, A. Modlińska, R. Dąbrowski, Opto-Electron. Rev. **18**, 63 (2010)
22. J. Świergiel, I. Płowaś, J. Jadżyn, J. Mol. Liq. **220**, 879 (2016)
23. E. DelRe, F.D. Mei, J. Parravicini, G. Parravicini, A. Agranat, C. Conti, Nat. Photonics **9**, 228 (2015)
24. J. Parravicini, E. DelRe, A. Agranat, G. Parravicini, Phys. Rev. B **93**, 094203 (2016)
25. J. Parravicini, E. DelRe, A. Agranat, G. Parravicini, Nanoscale **9**, 9572 (2017)
26. D.V. Matyushov, J. Chem. Phys. **145**, 034504 (2016)

27. G. Johari, J. Chem. Phys. **138**, 154503 (2013)
28. G. Johari, J. Chem. Phys. **139**, 026101 (2013)
29. G. Johari, J. Chem. Phys. **145**, 164502 (2016)
30. S. Samanta, R. Richert, J. Chem. Phys. **142**, 044504 (2015)
31. S. Samanta, R. Richert, J. Phys. Chem. B **120**, 7737 (2016)
32. V.V. Kochervinskii, S.A. Bedin, I.V. Razumovskaya, M.A. Shcherbina, M.I. Buzin, I.A. Malyshkina, A.M. Prazdnichnyi, A. Pavlov, O. Vasilevskii, Polym. Sci. Ser. A **58**, 345 (2016)
33. Y. Kurzweil-Segev, A.G. (Gutina), I. Popov, D. Golodnitsky, Y. Feldman, Phys. Chem. Chem. Phys. **18**, 10992 (2016)
34. F. Huang, C. Hu, H. Tian, et al., Cryst. Growth Des. **19**(9), 5362 (2019)
35. F. Huang, C. Hu, Z. Zhou, X. Meng, P. Tan, Y. Wang, C. Wang, H. Tian, Acta Mater. **200**, 24 (2020)
36. J. Parravicini, Phys. B **541**, 54 (2018)
37. R. Boyd, *Nonlinear Optics*, 3rd edn. (Academic Press, San Diego, 2008)
38. P. Power, *Fundamentals of Nonlinear Optics* (Taylor & Francis, Boca Raton, 2011)
39. R. Richert, J. Phys. Condens. Matter **29**, 363001 (2017)
40. D. Pierangeli, F.D. Mei, J. Parravicini, G. Parravicini, A.J. Agranat, C. Conti, E. DelRe, Opt. Mater. Express **4**, 1487 (2014)
41. M. de Oliveira, *Equilibrium Thermodynamics*, 2nd edn. (Springer, Heidelberg, 2017)
42. X. Huang, P. Tan, Y. Wang, et al., Cryst. Growth Des. **22**(5), 3485 (2022)
43. I. Panzic, A. Bafti, *Perovskite Ferroelectric/Relaxor Materials/Systems for Capacitive Energy Storage* (Springer International Publishing, Cham, 2025), pp. 95–130
44. M. Okutan, F. Yakuphanoglu, S.E. San, O. Koysal, Phys. B **368**, 308 (2005)
45. G.P. Johari, J. Chem. Phys. **112**, 7518 (2000)
46. G.P. Johari, J. Chem. Phys. **112**, 8958 (2000)

Experimental Techniques to Investigate the State-of-Order in Condensed Matter

Mauro Coduri, Leonardo Lo Presti, Giovanni Macetti, Luca Sironi, Margherita Vacchini, Raffaella Soave, Stefano Americo, and Marco Scavini

Abstract Defects are probably as important as the crystal structure to determine the macroscopic properties of crystalline materials. The performance of several cutting-edge applications, including ferroelectric relaxors and ionic conductors, critically depends on the nature and interplay of defects at different length scales. Unfortunately, the precise characterization of defects in terms of their number, quality, and structure is often a challenging task, because of their intrinsically statistical (entropy-driven) nature. Only in recent decades, impressive advances made in materials characterization and computational modeling have allowed access to unprecedented insight into the defective structure of crystals. This chapter is structured as a guided journey through the main state-of-the-art techniques for characterizing the state of order in solids. First, modern powder X-ray diffraction and total scattering methods are dissected. Then, Diffuse Scattering tools will be briefly illustrated, with reference to state-of-the-art Literature. Finally, we introduce

M. Coduri (✉)
Department of Chemistry, Università degli Studi di Pavia, Pavia, Italy
e-mail: mauro.coduri@unipv.it

L. Lo Presti · G. Macetti · M. Vacchini · M. Scavini
Department of Chemistry, Università degli Studi di Milano, Milano, Italy
e-mail: leonardo.lopresti@unimi.it; giovanni.macetti@unimi.it; margherita.vacchini@unimi.it; marco.scavini@unimi.it

L. Sironi
Department of Chemistry, Università degli Studi di Milano, Milano, Italy

Department of Chemistry, Durham University, Durham, UK
e-mail: luca.sironi1@unimi.it; luca.sironi@durham.ac.uk

R. Soave
CNR-SCITEC, Milano, Italy
e-mail: raffaella.soave@scitec.cnr.it

S. Americo
CNR-SCITEC, Milano, Italy

Nanochemistry, Istituto Italiano di Tecnologia, Genova, Italy
e-mail: stefano.americo@scitec.cnr.it; stefano.americo@iit.it

J. Parravicini (ed.), *Fröhlich Entropy Estimation and Dielectric Calorimetry*, Advances in Dielectrics, https://doi.org/10.1007/978-3-032-21454-6_4

the problem of defects in molecular materials, where internal degrees of freedom and chemical disorder add a further level of complexity, but also provide exciting opportunities to design new categories of innovative materials.

1 Introduction

Not a day goes by without dealing with crystals in some way. Crystalline materials enable the operation of our laptops, smartphones, cars, and home appliances. A crystal is a well-defined phase of matter, generally solid (although liquid crystals also exist), characterized by translational invariance at the atomic scale. This implies that atoms, ions, and molecules, each with nanometer-scale dimensions, are repeated uniformly in space to form an ordered array, which is called a lattice. Any pair of geometrical points that can be joined by an elementary translation or by some linear combinations of such translations sits on the so-called lattice nodes. A unit cell corresponds to the smallest translation-independent volume of the lattice. It is described by a set of geometric parameters, the unit cell edges and angles, which express the mutual length and orientations of the edges of the corresponding polyhedron. The unit cell is representative of the nature of the material, as different chemical compounds are invariably associated with different dimensions and shape of their unit cells. It is important to understand that the unit cell is an idealized model of the crystal, where the entire lattice is somehow averaged out into its translation-independent part.

Typically, the unit cell parameters are obtained by interpreting X-ray (or electron, neutron) diffraction patterns. Each diffraction effect contains information about the whole lattice that is exposed to the probe beam. Thus, the cell edge lengths estimated, for example, by X-ray crystallography, correspond to average elementary translations and cannot account for local deviations from the ideal crystallographic symmetry. Equivalently, it can be stated that the crystallographic cell edges catch the long-range translational order of ideally perfect lattices. In this context, the term "long-range" refers to the fact that the average unit translations are the same throughout the macroscopic volume of the crystal and correlate any pair of lattice nodes, irrespective of their distance. On the contrary, the term "local" refers to the neighborhood of any region of interest within the unit cell. Operatively, also the "local" structure comes from the average of all the same coordination sites in the crystal (see below), but the key point is that it may differ from the translation-invariant average crystal structure, as estimated by standard crystallographic and X-ray diffraction (XRD) methods. There are many possible reasons for this mismatch, all of which fall within the broadest category of so-called lattice defects.

It goes without saying that real crystals are not perfect. Whenever any symmetry-invariance constraint fails to be fulfilled, we say that the crystal bears a defect. The classification of defects spans quite a rich landscape, and the reader is encouraged to examine dedicated textbooks if they feel that they should delve deeper into the topic [1–3]. We briefly recall here the most important defect types (Fig. 1). Any misplaced atom in the lattice constitutes a point (zero-dimensional) defect (Fig. 1a,c,d,e),

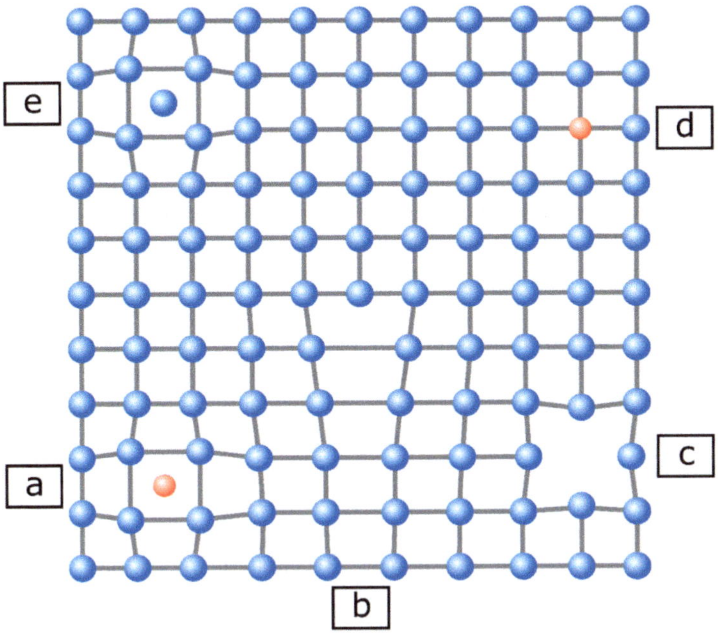

Fig. 1 Main structural defects in crystal lattices. (**a**) Interstitial. (**b**) Dislocation. (**c**) Vacancy. (**d**) Substitutional impurity. (**e**) Frenkel defect, i.e., an off-lattice atom or ion that occupies a vacant site nearby

which can be further classified as vacancy (missing atom, Fig. 1c), interstitial (Fig. 1a,e) or substitutional (Fig. 1d). Point defects are randomly distributed in most cases. However, since the very beginning of materials science, it has been recognized that defects can be structured in some way. A structured defect [4, 5] shows some degree of order, as for example it manifests preferentially in pairs [6], or it extends (quasi-)periodically throughout the lattice [7, 8].

From a broader perspective, structured defects are correlated, i.e., they exhibit some structural constraints that produce a coherent arrangement over large (long-range) spatial regions of the lattice. Thus, structured defects may amplify local (short-range) deviations, possibly affecting the material's properties up to the mesoscale (10^{-6} m) [9–11]. Higher-dimensional defects like dislocations (Fig. 1b) can be described as structured arrangements of point defects, either vacancies (Fig. 1c) or interstitial atoms (Fig. 1a). Dislocations are one-dimensional (1D) defects that arise typically due to a misalignment of lattice nodes across a slip plane. They may be further classified as edge, screw or mixed, depending on topology and folding of the defective line. Closed dislocation loops may also form, like in the well-known Frank-Read mechanism [12]; partial dislocations are associated with the formation of higher dimensional defects, like stacking faults and antiphase boundaries. A full treatment of the theory of dislocations and other high-order defects is out of scope; the interested reader is referred to existing excellent books and reviews [13–15]. Concerning the present discussion, the following points are worth noting.

First, even highly crystalline materials contain very large defect count on absolute grounds. For example, even the crystalline Cu used in electric wires may host up to $\sim 10^8$ metal vacancies per cm^3 at room temperature (RT), as expected for an equilibrium concentration given the defect formation energy [16].

Second, lattice defects have different origins. Most are intrinsic, that is, they form spontaneously during the nucleation and growth processes [17]. There is an unavoidable thermodynamic drive toward the formation of intrinsic defects, which can be traced back to the entropy gain of the defective lattice over any ideally perfect one (see below). Moreover, kinetics also plays a role. Several functional materials, including silicon and semiconductor oxide ceramics, are crystallized from the melt, or anyway annealed at high temperature. The formation of an ideally perfect crystal would require a reversible (i.e. infinitely slow) cooling, as each crystal component would need an infinite amount of time to reach its correct lattice site. Another important source of defects is crystal compliance [18]: whenever a sufficiently large stress is applied along some material-specific crystallographic directions, plastic deformation takes place and new defects are created. Interestingly, it is not even required that stress overcomes the yield point in most cases. For example, it has been known for a long time that moderate shear stress can cause the slip deformation of certain materials [19]. This is much more probable at high temperatures, where the components of the stiffness tensor are lower, or in molecular materials, where only relatively weak intermolecular forces dictate the crystal packing. Early experiments by Read showed that a measurable internal friction exists in crystals, which could be explained by the motion of dislocations [20]. Thus, not only do dislocations define the boundaries between slipped and unslipped regions of the crystal, but also explain how these change in response to external stress.

The third key point is that defects are always intertwined with crystal inner dynamics [21–23]. Defects perturb the long-range structure, increasing the lattice strain locally and, henceforth, the internal energy of the crystal. The crystal could release this strain as soon as it is allowed to, triggering dynamical effects. For example, vacancies are crucial to explain the solid-state diffusion of heavy atoms in metals, alloys and dielectrics [24, 25], as well as the diffusion of oxide (and other anions) on the surface [26], thus accounting for—among others—the catalytic activity of heterogeneous catalysts [27]. Another example is the so-called Zener relaxation [28], when external stress induces an inelastic reorientation of defect pairs. This generates a measurable excess strain, which is released with a characteristic relaxation time when the stress is removed, as the time it takes for the material to recover its original shape [29].

Finally, the defect structure of a solid material is intertwined with its entropy,

$$S = S_{perfect} + S_{defects} \tag{1}$$

$S_{perfect}$ is the intrinsic entropy content of the perfect (long-range ordered) lattice, which may be estimated by standard calorimetry measurements according to the second and third principles of thermodynamics. The $S_{defects}$ term specifically concerns the contribution of lattice defects, which exploit less (or not at all) long-range order. According to Boltzmann's definition, defect entropy measures the excess region of phase space that defects make accessible. Moreover, it can be decomposed into three main contributions:

$$S_{defects} = S_{con} + S_{vib} + S_{rot} \tag{2}$$

where S_{con}, S_{vib} and S_{rot} represent the corresponding configurational, vibrational and rotational components. Defects always increase the configurational entropy S_{con}, as they introduce additional atomic arrangements, which can be distributed in many ways through the lattice. This term also accounts for genuine chemical effects, like frustration of hydrogen bonds [30, 31] which, for example, explains the residual excess entropy of $\sim 3\,\mathrm{J{\cdot}K^{-1}{\cdot}mol^{-1}}$ at $0\,\mathrm{K}$ in crystalline hexagonal polymorph I of ice [32]. The vibrational term, S_{vib}, is also of paramount importance, as defects usually soften specific lattice vibrational modes. Therefore, the density of accessible phonon states is increased, as well as the corresponding entropy. The last term, S_{rot}, accounts for possible reorientation phenomena and it typically comes into play when the crystal contains molecular building blocks, i.e. when breaking of symmetry invariance of the molecular relative orientations introduces orientational (or conformational) disorder [33, 34]. When dealing with inorganic materials like metals, oxides and alloys, this latter term can be safely ignored in most cases. Statistical thermodynamics provides straightforward tools to compute all the three terms, if (classical or quantum) vibrational and rotational partition functions are known. However, when structured or correlated defects come into play, the relationship with entropy becomes more complex and, to the best of our knowledge, no universal recipes exist to predict $S_{defects}$ exactly. Rather, approximate expressions must be developed case by case. The reason is that structured defects exhibit spatial or energetic correlations that restrict their accessible phase space compared to isolated (random) point defects. Even though lattice entropy still increases due to the additional degrees of freedom, distinguishing the contributions from different defect sources becomes a hard business.

Entropy associated with defects contributes to the Gibbs free energy. Thus, it affects the stability of materials at different temperatures and is crucial to explain and rationalize temperature or pressure-induced phase transitions. Across a transition, thermodynamics-driven lattice modifications take place by exploiting the dynamic freedom granted by lattice defects and by the lattice distortions they create. This is particularly true in molecular materials, where soft phonon modes often define preferential reaction coordinates, which are useful to interpret displacive phase transitions [35]. For sure, detailed knowledge of nature and concentration of defects is mandatory to predict accurately the entropy of a real lattice and to understand its evolution across a phase transition.

A general conclusion might be summarized by a statement attributed to the French actor and former soccer player Éric Cantona: "*If someone is too perfect, they won't look good. Imperfection is important.*" Cantona's words could not be truer when applied to materials science. It has been understood since the 1960s that defects represent an extraordinary opportunity to boost the desired performance of a material. A modern branch of materials science concerns *defect engineering*, which aims at placing the correct amount and type of extrinsic defects in the desired crystalline matrix [36]. This is done often during the synthesis process, for example by adding specific reactants to the precursor mixture or by exposing the growing crystal to appropriate conditions, like rapid cooling or specific chemical environments [37–39]. The purpose is the precise tailoring of materials for targeted functionalities, with a growing space for applications in cutting-edge sustainable technologies. In most cases, these goals can be achieved by controlling the concentration and distribution of dislocation and grain boundaries, rather than by adding dopant species.

From the above discussion, it can be understood why increasing efforts were being made in the last decades to develop accurate and precise experimental probes to investigate defects in solids. Unfortunately, this is a very challenging task, owing to the statistical nature of the distribution of most defects in the crystal lattice. Probably, one of the first techniques that were developed to this purpose was differential dilatometry [40, 41]. The idea is to compare the macroscopic thermal expansion, i.e. the macroscopic volume change of the crystal as a function of temperature, with the microscopic one, as estimated from X-ray diffraction (XRD) experiments. If N is the number of unit cells of volume v in the lattice, the total volume of the crystal can be approximated as $V = N \cdot v$. If temperature is changed (usually raised well above RT), the new volume is $V+\Delta V = (N+\Delta N)\cdot(v+\Delta v) = N \cdot v + N \cdot \Delta v + v \cdot \Delta N$, provided that higher-order terms $\Delta N \cdot \Delta v$ are ignored. Here, N is the change of unit cell count due to the thermally induced release of free volume-type defects. Thus, it can be negative (positive) if interstitial (vacancy) defects prevail in the original sample. The expression for $V + \Delta V$ rearranges to $\Delta V = N \cdot \Delta v + v \cdot \Delta N$, which can be further simplified by dividing both terms by $V = N \cdot v$ to have the corresponding relative quantities. The outcome is

$$\Delta V/V = (\Delta v/v) + (\Delta N/N) \tag{3}$$

Thus, any difference in the macroscopic and microscopic relative thermal expansions can be explained in terms of competing defects. Unfortunately, this technique alone cannot provide the full picture, as for example it cannot distinguish between structured and random defects on absolute grounds. Moreover, it cannot be used to study systems where volume defects are equally distributed between interstitial-like and vacancy-like.

Nowadays, experimental approaches to study defects in solids are multidisciplinary and include the interplay of calorimetric, spectroscopic, microscopic and diffractometric approaches. For example, Transmission Electron Microscopy (TEM) provides atomic-scale imaging and allows to analyze the distribution of dislocations and grain boundaries [42, 43]. Scanning Electron Microscopy (SEM), especially if coupled with Electron Backscatter Diffraction (EBSD) and Energy Dispersive X-ray Spectroscopy (EDS), allows to map the microstructure of micrometer-large samples, and determines crucial microstructural information like grain size, texture and twin boundaries [43, 44]. Scanning Tunneling Microscopy (STM), which can be applied only to conductors and semiconductors, provides atom-scale surface imaging and local electronic property measurements, allowing to literally see individual defects [45]. Positron Annihilation Spectroscopy (PAS) was recently developed as a non-destructive probe to map low concentrations of defects in solids with almost atomic resolution [46]. The idea is to probe the materials with a beam of positrons and to detect the characteristic photons that result from the electron-positron annihilation process. Defects of different nature influence the electronic properties of their local environment, such as density of states, momentum distribution and formal oxidation state. The net outcome is a detectable chemical shift in the energy of the outgoing photons. This way, different classes of defects can be mapped throughout the sample volume.

Computational methods often stand alongside experimental characterizations, providing models to explain and disentangle complex outcomes from different techniques. The interplay of experiments and modeling approaches allows one to gain information about lattice defects at different length scales, granting unprecedented insights into the inner lattice structure and dynamics. Of relevance in this context are methods based on dielectric spectroscopy and Frölich entropy, which will be dissected in other chapters.[1] Here, we will focus mainly on advanced XRD tool based on synchrotron light sources, that will be compared to X-ray absorption spectroscopy, with applications on perovskites and other functional materials. A discussion on the capabilities of single-crystal XRD techniques, coupled with Molecular Dynamics, will close the chapter, demonstrating how the interplay of experimental and computational techniques can disentangle configurational and thermal disorder in complex materials and molecular crystals.

2 Disorder from X-Ray Powder Diffraction

Powder diffraction is probably the first-choice method to gain insights into structural disorder. The advantage of powder diffraction lies in its availability in materials chemistry laboratories, as well as in the relative easiness to produce the specimen compared with single crystal techniques (see below). In a powder diffraction experiment, it is assumed that all crystallites are randomly oriented. The resulting diffraction pattern is obtained by recording the scattering intensity as a function of the diffraction angle. Structural disorder can impact the powder diffraction pattern in three main ways: (1) peak broadening, (2) increased atomic displacement

[1] See in particular chapters "Introduction: Dielectric Calorimetry and Fröhlich Entropy Estimation", "Meaning and Applications of Fröhlich Entropy", "Fröhlich Entropy of Nanoparticles: Investigation of Phases, Phase Transitions and Melting", "Dielectric Thermodynamics of Organic Semiconductors", and "Order–Disorder Transitions of Ferroelectric Perovskites by Fröhlich Entropy Investigations".

parameters (ADPs), and (3) diffuse scattering (DS). Below, these three effects are described in detail.

Peak broadening in a diffraction signal reflects a limited coherence length in the direction perpendicular to the corresponding lattice plane. This broadening is often associated with the extended defects mentioned in Sect. 1 above. The most intuitive cause of peak broadening is the finite size of the crystals: smaller crystals typically lead to broader diffraction peaks. The first equation used to estimate crystal size from XRD was developed by Scherrer in 1919 [47]. Although this method relies on several approximations (e.g., monodisperse distribution, spherical shape, and the absence of other defects), it is commonly used for a quick estimate, particularly when dealing with very small nanoparticles. However, other defects can also contribute to XRD peak broadening. In 1953, Williamson and Hall (WH) examined XRD line profiles of cold-worked metals and identified two key contributions to broadening, based on their different dependencies on the scattering angle: crystal size and microstrain [48]. Microstrain refers to defects that cause a distribution of interplanar distances on the micron scale, leading to peak broadening. In the case of worked metals, this microstrain is often caused by dislocations (see Fig. 1b). While it is possible to simulate the effect of dislocations on an XRD pattern and predict their formation in simple systems, accurately determining dislocation density is challenging [49], as other defects can also contribute to peak broadening. For solid solutions, microstrain can arise from uneven distribution of atoms or ions at the nanometer scale, especially when the ions involved differ in size. In this case, unit cells may have different compositions and therefore different cell parameters. As a result, the Bragg peaks from each set of unit cells will occur at slightly different d-spacings, contributing to peak broadening.

While Scherrer and WH methods are commonly used to estimate crystal size and microstrain, they are empirical and based on several important approximations, such as the absence of other defects, spherical monodisperse crystals, and a Lorentzian and/or Gaussian peak shape. Nowadays, analytical functions have been developed to relate various types and densities of defects to the magnitude of peak broadening and its dependence on Miller indices [50, 51]. For instance, dislocation density, which is particularly relevant in metals and alloys due to its link to plastic deformation, can be estimated by powder diffraction measurements [52], even for ceramics [53]. Another important planar defect is the stacking fault, which is common in single metals [54] but also plays a significant role in semiconductors [55] and layered structures, where abnormal layer stacking [56] can affect functional properties and is probed by a peculiar broadening of XRD reflections.

Another relevant extended defect is the antiphase boundary, which occurs when a material undergoes an order/disorder transition as a function of temperature, as seen in the Cu_3Au alloy [57]. In the disordered state, two elements distribute randomly in the same crystallographic site. However, in the ordered state (below a critical temperature), the elements order and occupy different sites, lowering the symmetry of the system. The ordered phase nucleates locally in various parts of the crystal and grows. When these regions meet, they can form a defect layer at the interface, which affects only the broadening of the superstructure reflections.

Now, consider the effect of point defects (Fig. 1). In the case of isovalent doping—replacing an atom or ion with one of the same valence—the substitution typically does not cause significant structural changes in the surrounding atoms or ions. For example, when Ni^{2+} ions replace Mg^{2+} ions in MgO, both ions have the same valence and are fully soluble. Ideally, this substitution should not affect the atomic arrangement. However, Ni^{2+} ions are smaller than Mg^{2+} ions [58], and they tend to maintain the same bond distance with oxygen as in NiO (~2.09 Å), which is shorter than the Mg-O bond in the host structure (~2.11 Å). As a result, Ni^{2+} ions attract O^{2-} ions, leading to a local distortion around the dopant. The disorder induced by isovalent doping is generally small, also because when the size mismatch is significant, the dopant ion may not be soluble, leading to the formation of a secondary phase. Larger structural distortions are typically observed in aliovalent doping, where an ion is replaced by one with a different valence. This is commonly seen in oxide ion conductors such as ZrO_2 or CeO_2. When doped with trivalent cations, the system maintains electroneutrality by forming oxygen vacancies-in this case, one for every two dopant ions [59]. Doping affects both the cation and anion substructure. The local disorder arises not only from the size mismatch between ions but also, and more significantly, from the oxygen vacancies. The removal of a doubly charged ion causes a strong local relaxation, attracting nearby oxygen ions and repelling cations [60].

The degree of disorder in a material can be assessed through powder diffraction. In a perfectly ordered crystal, where the atomic arrangement is ideally defect-free, the content of each unit cell is identical. However, in the presence of defects, the actual positions of the ions may vary, depending on the nature and location of the defects within each unit cell. On average, though, the positions of the ions remain similar to those in the defect-free material, with only minor adjustments in lattice parameters or atomic coordinates. Despite this, defects can significantly influence the XRD pattern. X-rays are scattered by the electrons of the atoms or ions in the material, and although the average position of the ions may remain the same as in a defect-free material, in a defective material, ions will occupy different positions across the unit cells. This scattering is spread over a larger region than the one occupied in each unit cell, which is akin to a very large atomic vibration, leading to a noticeable intensity decay at higher scattering angles in the XRD pattern.

For XRD, structural disorder and atomic vibrations have similar effects (see also Sect. 6 below). These two factors can be distinguished by observing the variation of the ADPs of different ions with temperature. In the absence of phase transitions, structural disorder remains constant (referred to as "static" disorder), while atomic vibrations intensify with temperature above a certain threshold [62]. For instance, this method has been used to analyze disorder in yttrium-doped ceria at both anion and cation sites [63]. Notably, in the presence of static disorder, the variation in ADPs can be significant, potentially spanning an order of magnitude [64]. When symmetry allows, static disorder may occur in a specific direction, such as the off-centering of a cation within an octahedron in perovskites or the relaxation of an oxide ion towards an oxygen vacancy [61]. Directional static disorder can be analyzed using anisotropic ADPs, which describe the ADP as a tensor. The tensor parameters are calculated by modeling the intensity decay as a function of scattering angle and Miller indices. As a result, a sufficient number of reflections is necessary for meaningful modeling. This approach is reliable only when the ADP is predominantly influenced by static disorder, as the contribution from anisotropic vibrations is generally small in ionic compounds. Moreover, since X-ray scattering depends on scattering factors and the number of electrons, reliable ADPs, especially anisotropic ones, can typically only be assessed for cations, especially heavy ones. This makes neutron diffraction necessary to study oxygen ions and lighter elements. Anisotropic ADPs also require a greater number of reflections, a high intensity-to-noise ratio, and ideally, good instrument resolution to resolve reflections with similar d-spaces. These conditions generally necessitate synchrotron radiation.

It is important to remember that static disorder has a physical interpretation: it represents an additional contribution, in respect to atomic vibrations, to the mean square displacement (msd), defining the amplitude of an atom or ion displacement (in Å^2) from its equilibrium position [62, 65]. In disordered materials, especially at low T, ADPs are dominated by the static disorder contribution. As displayed in Fig. 2d–f, when ADPs are evaluated in disordered systems such as solid solutions, the comparison with the parent compounds allows the quantification of static disorder as a function of composition and/or temperature. Finally, if static disorder is excessively high, it could indicate an inaccurate structural model, such as incorrectly positioning an ion in a special site when it does not belong there [66–68].

The analysis of peak broadening, as well as of ADPs, provide some hints on the nature of disorder. However, owing the properties of the Fourier transform that relates any real crystal with its reciprocal space image, any deviation from the long-range structure in the sense explained in Sect. 1 affects the whole diffraction output. This implies that disorder has its own signature in the so-called diffuse scattering (DS) features, which are detailed in the next section.

3 Diffuse Scattering

In general, the position of diffraction effects in reciprocal space can be explained in terms of the Bragg Law

$$n\lambda = 2\, d_{hkl} \sin\theta \tag{4}$$

where θ is half the angle between the incident and diffracted X-rays, λ the wavelength of X-rays, and hkl are the integer crystallographic indices of parallel real-space planes with interplanar distance d_{hkl} [69]. Disorder, either static or dynamic, breaks the ideal translational invariance of the crystal and reduces the correlation among symmetry-dependent atoms. The outcome is a general broadening of diffraction reflections, which can be easily quantified by accurate X-ray powder diffraction experiments and is clearly detectable already in single crystal diffraction images. The effect is more prominent for high-order (high θ) data, where Bragg

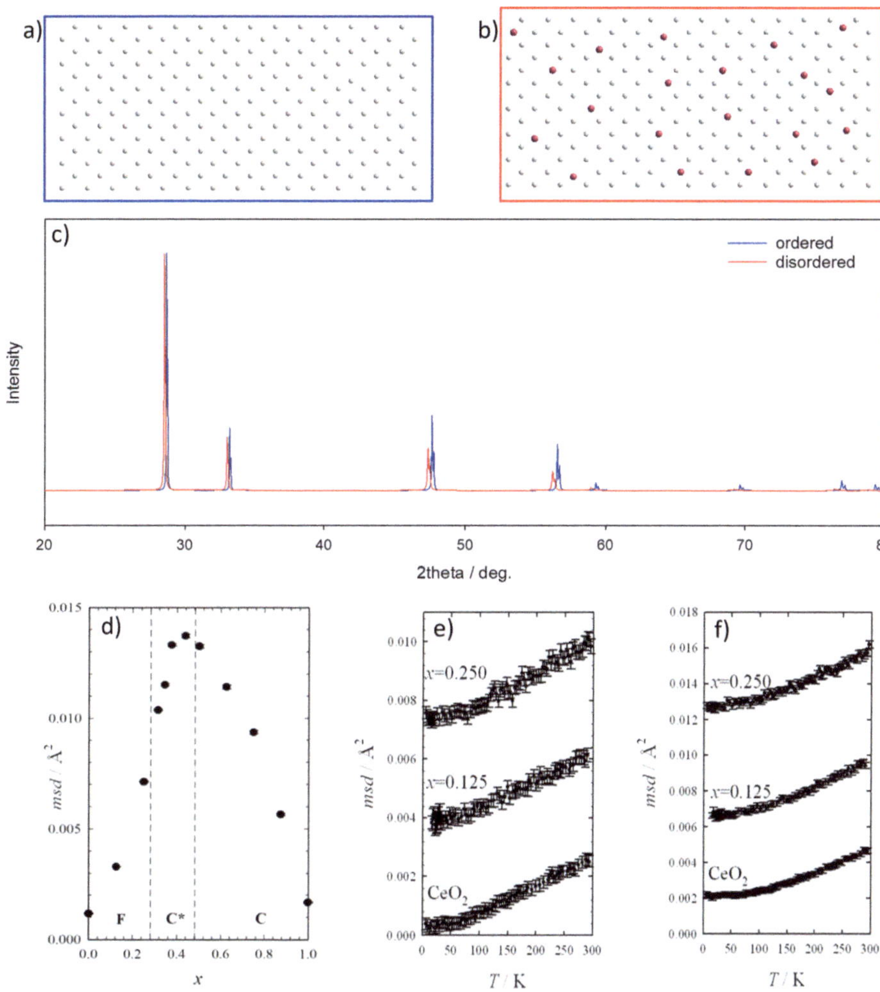

Fig. 2 (**a–b**) Pictorial view of an ordered fcc atomic arrangement (**a**) and a disordered arrangement where 10% of atoms are substituted by a larger dopant, which occupies a slightly different position than the atoms in the host matrix. (**c**) corresponding simulated XRD patterns. (**d–f**) Experimental evidence of static disorder from XRD on Y-doped ceria compounds. (**d**) average ADPs of all ions across the whole CeO_2-Y_2O_3 compositional range at constant temperature, (**e–f**) evolution of ADPs as a function of temperature for the cations and oxygens sites, respectively. Adapted, with permissions, from [61]. (Courtesy of the original publisher ©)

intensities are smaller, and the scattering is dominated by internal ("core") electrons [70]. As the total scattering power must be conserved, the missing intensity at Bragg positions is redistributed across the reciprocal space, producing what is known as diffuse scattering (DS). The theory of diffuse scattering has been well developed since the early days of solid-state physics, and this is not the correct place to discuss it in detail. Excellent reviews and textbooks exist, to which the reader interested in deepening the theoretical aspects is referred [71, 72]. Here, we limit ourselves to briefly recap the main tenets of the model.

The first and ubiquitarian source of DS is lattice dynamics, including atomic thermal motion involving the normal vibration modes of the molecular structure, if any, and lattice phonons. Together, molecular modes and phonons give rise to the so-called thermal diffuse scattering (TDS). Phonons are collective lattice vibration modes that can be understood as lattice excitations, which in turn can be modeled as wave-like quasiparticles with their own dynamics. Phonons manifest a dispersive behavior, in the sense that the frequency ω_j of each j-th phonon mode is a continuous function of its wavevector $\mathbf{q} = (2\pi/\lambda)\,\hat{q}$. Thus, their dynamic properties, like kinetic energy, velocity, amplitude and so on, depend on both the crystal momentum $\hbar\mathbf{q}$ and the direction of propagation $\hat{q} = \mathbf{q}/|\mathbf{q}|$ in the reciprocal space. Each curve $\omega_j(\mathbf{q})$ is the j-th "branch" of the overall dispersion relations.

Analogously to real particles, phonon modes can act as scattering sources for neutrons, electrons or X-ray photons. The main contribution to TDS comes from the one-phonon scattering, which is also referred to as first-order TDS. If needed, higher-order corrections, corresponding to two-(or more) phonon events, can be also accounted for. Under the validity of the harmonic and adiabatic approximations, for each wavevector $\mathbf{q}$, the one-phonon differential cross section at each reciprocal space point $\mathbf{Q}$ is proportional to [72]:

$$\left(\frac{d\sigma}{d\Omega}\right)^{(1)} \propto \sum_j \frac{E_j(\mathbf{q}) \cdot |G(\mathbf{Q}, \mathbf{q})|^2}{\omega_j^2(\mathbf{q})} \cdot \sum_h \delta(\mathbf{Q} \pm \mathbf{q} - \mathbf{Q_h}) \tag{5}$$

Equation (5) expresses the probability that X-rays are scattered by a phonon into the solid angle $d\Omega$ in units of area per solid angle. The larger the cross section, the higher the corresponding scattered intensity. Here $\mathbf{Q} = \mathbf{k} - \mathbf{k_0}$ is the scattering vector, expressed as the difference between the wavevectors of the scattered ($\mathbf{k}$) and incident ($\mathbf{k_0}$) beams, and $\mathbf{Q_h}$ is the reciprocal lattice vector of the node h with integer coordinates h, k, l. Other relevant quantities in (5) are $E_j(\mathbf{q})$, the energy of the j-th phonon branch at $\mathbf{q}$, and $|G(\mathbf{Q}, \mathbf{q})|$, the one-phonon structure factor amplitude. The latter accounts for the scattering power of each phonon branch, which also depends on the polarization state of the quasiparticle.

Equation (5) is important for two reasons. First, the summation over a series of Dirac δ functions restraints the relevant scattering vectors to $\mathbf{Q} = \mathbf{Q_h} \pm \mathbf{q}$. However, a huge number of $\mathbf{q}$ vectors exist, which are uniformly distributed across the symmetry-independent part of the reciprocal cell. Thus, they give rise to a quasi-continuum of allowed $\mathbf{Q}$ vectors and result is a diffuse first-order TDS intensity in reciprocal space points $\mathbf{Q}$, not necessarily close to Bragg peaks at $\mathbf{Q_h}$. Second, it is known that low-$\mathbf{q}$ phonons of the acoustic branches fulfill the condition:

$$\lim_{\mathbf{q}\to\mathbf{0}} \omega_j(\mathbf{q}) = 0 \tag{6}$$

Low-frequency acoustic modes thus produce peaks of scattered radiation very close to Bragg positions ($\mathbf{Q} \rightarrow \mathbf{Q_h}$) and, due to the ω^{-2} factor in (5), their contributions to the TDS intensity are significant. The reason is that low-frequency phonons imply high-wavelength collective oscillations which may involve translation-related atoms or molecules. Thus, the correlation between distant atoms that is lost due to the high-frequency vibrations (see above) is—at least partially—recovered. The net effect is that acoustic branches of TDS generally produce an overestimation of the measured intensity at Bragg positions $\mathbf{Q_h}$; in X-ray experiments, the error could be significant for weak high-θ reflections, leading to inaccurate structural models for atomic and molecular geometries and, in general, making comparisons with theoretical predictions more difficult [73].

A key point is that disorder is an important source of diffuse scattering as well. Any failure to fulfill translational invariance implies the reduction of the Bragg scattering amplitude at the reciprocal lattice vectors $\mathbf{Q_h}$. The net result is that static disorder in an XRD signal extends beyond just an increase in ADPs. As mentioned above, the intensity lost from the Bragg peaks accumulates as DS in the background [74, 75]. A common example of DS is found in amorphous materials like glass: the absence of Bragg peaks does not mean a complete lack of structural ordering. While long-range order is absent, ions still maintain regular coordination environments (such as tetrahedral or octahedral coordination). Although diffraction peaks are missing, the XRD pattern exhibits broad bumps rather than being flat. Generally, any deviation from a perfectly ordered crystalline structure results in DS, which is typically located in the background between Bragg peaks, if present.

Any missing Bragg intensity is never lost but contributes to DS components, which can be investigated *quantitatively* if one has access to the whole reciprocal space. This is possible by performing a single crystal XRD study, which however might be—and in fact often is—a hard business. The reason is that it is difficult to obtain even μm-size single crystals, with the required quality [76]. Most chalcogenides and halogenides form twins while they are growing, and the resulting diffraction image is the superposition of several differently oriented lattices [77, 78]. In general, high-quality powdered samples are easier to synthesize, and this is the reason why DS is often tackled by means of X-ray powder diffraction. However, significant challenges arise, which are related to modeling and interpreting the DS contributions to the observed XRD signals. The reason is that in powder XRD methods, the whole reciprocal space is necessarily projected onto a 1D pattern (intensity vs. 2θ). If the crystalline structure of a material is known, and a reasonable guess can be provided for local distortions, the corresponding diffraction pattern (Bragg + DS) can be simulated. However, accurately modeling *only* the DS in a crystalline material requires (1) an exceptionally high signal and (2) isolating the DS from all other diffracted signals, which is a difficult and often impractical task experimentally [79]. Therefore, DS is typically studied alongside Bragg scattering.

Another limitation of studying DS in projected reciprocal space (d_{hkl}, 2θ) is that it assumes very large features, which complicates the differentiation of various models. A noticeable exception is the use of the Debye Scattering Equation (DSE) method, which uses both Bragg peaks and DS to describe, as an example, the (defective) atomistic structure and the morphology of nanoparticles [80, 81]. This approach necessitates synchrotron radiation to gather data with sufficient instrument resolution and signal-to-noise ratio to record reliable value of Bragg and DS, determined by appropriate background subtractions.

4 Disorder in Direct Space: The Pair Distribution Function

The complexity in retrieving structural information from DS in the reciprocal space calls for an alternative approach: the Pair Distribution Function (PDF). The PDF is the Fourier Transform (FT) of the diffraction signal from the specimen and gives the probability of finding an atom pair as a function of the interatomic distance r. Therefore, peaks in the PDF correspond to interatomic distances. Since the FT is applied to the entire scattered pattern (after appropriate corrections), including both Bragg and DS, the PDF represents a form of total scattering, like DSE. Originally developed only for studying liquids and amorphous materials [82], the PDF technique has been applied to crystalline materials in recent decades, due to (1) the development of free, user-friendly software like PDFgetX2, PDFgetX3 [83], and PDFgui [84], and (2) advancements in the quality and speed of data acquisition from synchrotron and neutron sources [85]. Reliable PDF data cannot be obtained from conventional laboratory diffractometers with Cu radiation. Specifically, the PDF is defined as the sine FT of the normalized diffraction signal from the specimen:

$$\mathrm{PDF}(r) = \frac{2}{\pi} \int_0^{\infty} F(Q) \sin(Qr)\, dQ \tag{7}$$

Where $F(Q)$ is the diffraction signal from the specimen after proper normalization, and Q is the modulus of the scattering vector, defined as:

$$Q = \frac{4\pi \sin\theta}{\lambda} \tag{8}$$

The integral in the equation ideally extends to infinite Q, which is practically impossible, as it would involve an infinitesimally small wavelength, equating to infinite energy. Truncating the integral to a finite Q value introduces two main issues: (1) it creates spurious oscillations (known as truncation ripples) at short interatomic distances, and (2) it reduces the resolution in r [86]. As a result, laboratory diffractometers with Cu radiation typically have a Q_{max} value of about 8 Å^{-1}, leading to poor quality data. Reliable Q_{max} values are generally achieved at around 25–30 Å^{-1}, which is typically possible using synchrotron radiation, where the incident wavelength can be tuned to optimize both the signal and the Q-range. An advantage of the PDF approach is that, unlike the blurred signals of DS, it provides direct information about interatomic distances, which can be compared to reference values.

The most practical method for extracting structural information from an experimental PDF curve is through *direct analysis*. Each PDF peak corresponds to a specific atom pair (or a combination of more pairs), with the peak position reflecting the interatomic distance, peak broadening indicating the amplitude of vibration, and peak intensity being dependent on the multiplicity and the types of atoms involved. However, the absolute values of the PDF are also influenced by the notation used. Due to the availability of user-friendly software, $G(r)$ has become the most commonly used notation for this analysis. Given a structural model, $G(r)$ can be calculated as follows:

$$G(r)_{calc} = \frac{1}{r} \sum_{\nu} \sum_{\mu} \left[\frac{b_\nu b_\mu}{\bar{b}^2} \delta(r - r_{\mu\nu}) \right] - 4\pi\rho_0 r \tag{9}$$

Where r is the interatomic distance, b is the scattering length (related to form factors for X-rays), and ρ_0 is the number density (atoms per unit volume). To account for atomic vibrations, the delta function in the equation is replaced by a Gaussian function. As a result, the PDF is most sensitive to heavier elements. For example (Fig. 3a), in cerium dioxide, the shortest interatomic distance corresponds to the Ce-O pair at approximately 2.34 Å (with a multiplicity of 8), followed by 6 O-O pairs and 12 Ce-Ce and 12 O-O pairs at 3.80 Å.

Although three PDF peaks would be expected within 4 Å distance, only two are observed. This is because the experimental and simulated $G(r)$ for X-rays are primarily sensitive to the Ce-Ce atom pairs, while the Ce-O atom pairs produce peaks with lower amplitudes, and the O-O pairs, if resolved, have negligible intensity [63]. Additionally, termination ripples due to a limited Q_{max} mostly affect the peaks at shorter interatomic distances, which can complicate the description of metal-oxygen pairs in ceramics. This issue is also present in other functional materials, such as lanthanide-based perovskites (e.g., $BaCeO_3$) [87]. In this framework, neutrons PDF provides a different contrast between elements, as the scattering lengths b in Eq. (9) are independent from the scattering form factors.

Unlike other experimental techniques, the PDF approach offers the advantage of providing structural information beyond just the first coordination shell. Since the PDF signal is the FT of the diffraction intensity scattered by the material, the maximum range of interatomic distances that can be probed is limited only by the instrumental resolution. A limited instrumental resolution results in a drop in intensity as a function of r [88]. Therefore, when diffraction data are collected with an instrument offering sufficient resolution, PDF data can extend to several nanometers, enabling structural investigation across various length scales. Therefore, when the instrument resolution is large enough, the extent of the PDF signal is dictated by the structural coherence, as shown in Fig. 3b, where the PDF signal of micro- and nanoparticles is compared.

To refine a structural model using PDF data within a specific interatomic distance range, a method called *real space Rietveld* has been developed, analogous to the conventional Rietveld method [89]. This approach allows for the quantitative

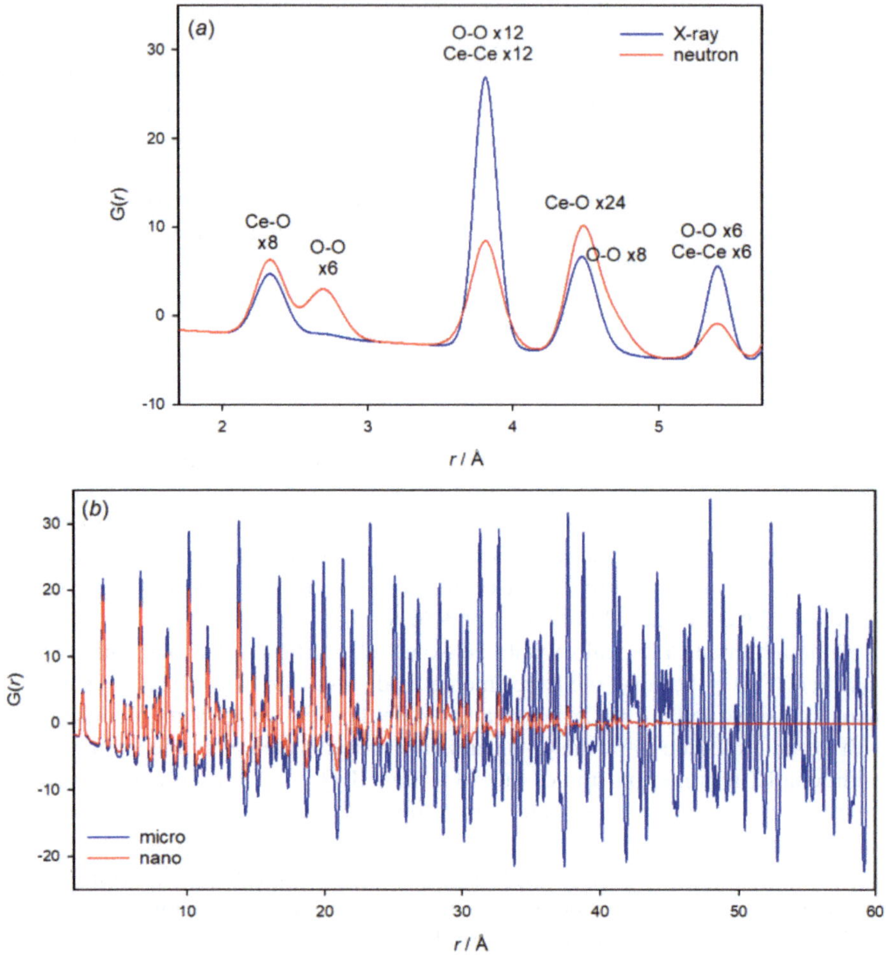

Fig. 3 (**a**) Simulated X-ray (blue) and neutron (red) PDF curves with the $G(r)$ formalism for CeO_2, using a 5.41 Å, msd(Ce) = 0.003 Å^2 and msd(O) = 0.007 Å^2. (**b**) Simulated X-ray PDF for CeO_2 with micrometer crystal size (blue) and 5 nm crystals (red)

determination of crystallographic parameters (such as cell parameters, atomic coordinates, and ADPs). By minimizing the difference between experimental and calculated PDFs, this method can be used to distinguish between different structural models. It can also be applied to different ranges of interatomic distances, facilitating a multi-scale structural analysis.

In disordered systems, two types of atomic structures must be defined: the *average structure* and the *local structure*. The average structure is the one determined through Rietveld refinement (or single-crystal analysis, when available). It represents the average position of all atoms within the unit cells across the entire material. The local structure, on the other hand, is defined by the sum of all interatomic distances around each atom, integrated over all atoms in the sample. When a local distortion occurs, such as a "simple" ion relaxation towards or away from point defects or a Jahn-Teller (JT) distortion, it can result in a symmetry reduction compared to the average structure.

For example, a JT distortion induces the elongation of a regular octahedral environment, i.e. an atom M with six equivalent M-O distances. The key factor distinguishing local and average structures lies in the ordering of defects over long distances: if the JT distortion occurs randomly, the octahedra elongate randomly in space introducing local disorder while on average they appear to be regular. This is the case for $LaMnO_3$ at high temperatures, where the distortion is only local. If the JT distortion propagates over a long range, it affects the average structure as well [90].

In this context, the PDF method can describe both local and average structures. When the instrumental resolution allows, the structure determined by the PDF at larger *r* ranges (typically above 2–3 nm) progressively merge into the average crystallographic structure obtained from traditional Rietveld refinement or single-crystal analysis. In the absence of significant structural disorder, the local and average structures should be similar. However, in the presence of complex defect architectures, the PDF provides a unique opportunity for a multi-scale structural investigation. This is typically performed using a "box-car" refinement approach, where the same structural model is applied to different *r* ranges [64, 88]. This method is particularly effective in revealing domain structures, where defect correlations extend beyond the first coordination shells. Examples of such a disorder in material chemistry often arise at temperatures close to phase transition and/or in solid solutions with a large amount of substitutional ions, which are not likely to distribute ideally within the host matrix.

The case of rare earth (Gd, Sm and Y) doped ceria compounds is a typical example that matches together the different evidence of structural disorder described above. These materials are oxygen ion conductors of interest for use as electrolytes in solid oxide fuel cells. Doping trivalent cations in place of tetravalent Ce ions creates oxygen vacancies in the fluorite structure of CeO_2, enabling oxygen ion migration via a thermally activated hopping mechanism. However, the solid solution is far from ideal, as dopant ions and oxygen vacancies tend to cluster together. At high dopant concentrations (still within the solubility limit of the dopant in CeO_2), these clusters grow into nanodomains embedded in the ceria matrix. This domain structure affects the material's ionic conductivity, as the domains trap oxygen vacancies, making them unavailable for oxygen ion migration. Strong structural disorder is indicated by unusually large ADPs from XRD, while box-car refinements of PDF data reveal lower symmetry extending across several nanometers. When the dopant concentration reaches a critical percolation threshold ($\sim$30%), the domains coalesce, leading to a transformation of the average structure from fluorite to bixbyite, the native structure of the dopant oxide. Once the structure transforms to bixbyite, the underlying domain structure can be detected in the XRD pattern by broadening of the superstructure peaks characteristic of the bixbyite phase.

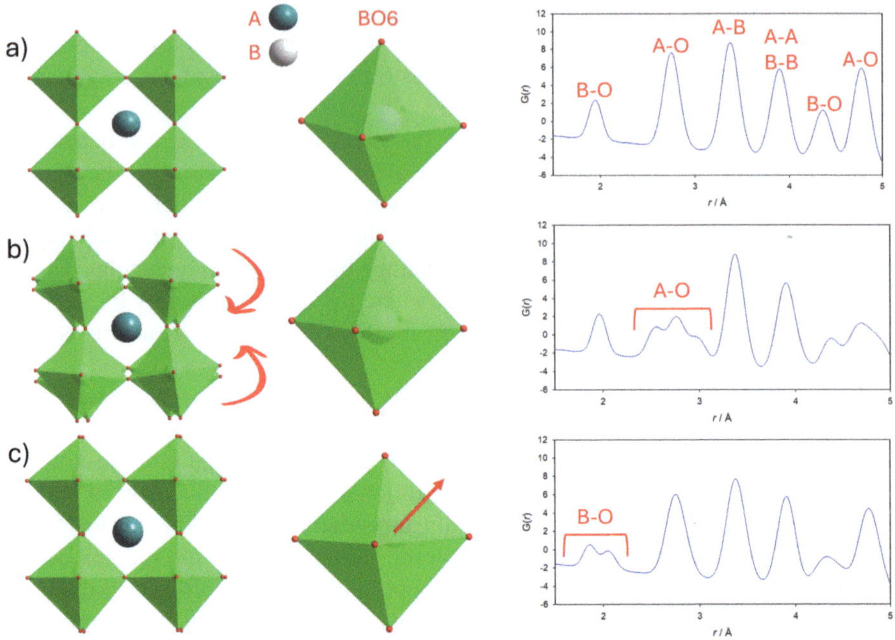

Fig. 4 Sketch of a perovskite octahedra arrangement in the case of a (**a**) cubic *Pm-3m*, (**b**) rhombohedral *R3m*, and (**c**) tetragonal *I4/mcm* symmetry. The left part highlights the A cations surrounded by eight BO_6 octahedra. The single octahedra are highlighted in the middle column. The right-hand part of the figure shows the corresponding PDF curves. Note that the distortions are maximized for improve readability

This broadening is caused by antiphase boundaries, which occur where dopant-rich nanodomains meet [63, 64, 91].

Perovskites are another structural system that is particularly suited for hosting extended structural disorder. In the simplest form, perovskite oxides with the formula ABO_3 are characterized by a three-dimensional network of corner-sharing BO_6 octahedra. The case of regular undistorted octahedra arranging with a cubic symmetry (e.g. *Pm*-3*m*) is shown in Fig. 4a. The B cation is located exactly at the center of the octahedron. Perovskites often undergo a variety of phase transitions as a function of temperature, involving a combination of octahedral tilting and ion displacement, while still maintaining the overall three-dimensional octahedral connectivity. These transformations significantly influence the dielectric properties.

In the field of perovskites, the PDF is exploited especially to determine local symmetry breaking, which reflect a different distribution of interatomic distances. A typical example is the off-centering of the B cation from the center of the octahedron, which occurs across a paraelectric to ferroelectric (FR) phase transition. As an example, in $BaTiO_3$ it corresponds to a series of phase transitions upon cooling: $Pm\bar{3}m \rightarrow P4mm \rightarrow Amm2 \rightarrow R3m$ [92]. The cubic phase is paraelectric while the others are FR. The structure of the rhombohedral phase is shown in Fig. 4b and the red arrow indicates the direction of the shift of the Ti ion. The spontaneous polarization is due to an uneven charge distribution within the octahedron, creating the potential for a ferroelectric material. Another common perovskite distortion is related to the tilting of TiO_6 octahedra along one or more axes. As an example, the antiferrodistorsive (AFD) tilting along the c axis drives to a tetragonal phase, space group $I4/mcm$, see Fig. 4c. This cubic-to-tetragonal distortion does not involve any off-centering of the B-cations, which maintains, therefore, a regular environment as in the parent cubic compound. A local symmetry breaking can be spotted through the PDF as a split of interatomic distances. In the case of the FR rhombohedral distortion described above, the split involves the first B-O distances, as the off-centering of the B cations create shorter and longer B-O distances within the octahedra. In the case of a AFD tetragonal distortion, the octahedra remain regular, but their tilting redistribute the distances between the A cation of the O ions. The effect of the distortion on the PDF profiles are displayed in Fig. 4.

When $BaTiO_3$ undergoes a cubic to tetragonal (e.g. $P4mm$) phase transition below 393 K, resulting in spontaneous polarization at room temperature, these interatomic distance modifications appear in the average structure. However, when $BaTiO_3$ is in nanoparticle form, this polarization is suppressed [93]. As mentioned in Sect. 2, nanocrystals experience substantial broadening of XRD peaks, which can obscure subtle structural distortions, such as peak splitting or small superstructure peaks. In this context, the real-space resolution of the PDF is independent of crystal size, making it an ideal technique for studying nanomaterials. In $BaTiO_3$, the PDF revealed a local tetragonal structure, suggesting that spontaneous polarization could potentially occur. However, the distortion is subtle and confined to the local scale (approximately 1.5 nm), meaning it is not strong enough to induce macroscopic spontaneous polarization.

In dielectric materials, the PDF is particularly useful for investigating relaxors, which display different temperature-dependent behaviour compared to conventional ferroelectrics [94]. Unlike ferroelectrics, relaxors do not exhibit a macroscopic phase transformation and instead show a smeared maximum in their dielectric permittivity rather than a sharp peak [95]. These characteristics are now believed to be linked to the formation of polar nano regions (PNRs)—randomly oriented polarized nanodomains that tend to align below a critical temperature. While no long-range phase transformation is observed, these polarized nanodomains are central to the unique dielectric properties of relaxor materials. Proving their existence using conventional crystallographic techniques is challenging, except through strong static disorder and DS. The PDF has been successfully used to quantify the fraction and size of PNRs in $Pb(Mg_{1/3}Nb_{2/3})O_3$ relaxors, building a model with low-symmetry domains (PNRs) within a cubic paraelectric matrix to match experimental PDF data in the nanometer range [96]. Jeong et al. showed that below a critical temperature, the fraction of PNRs increases until a percolation threshold is reached, which is consistent with the macroscopic polarization observed.

The PDF can also provide valuable structural information when an electric field is applied. By using a two-dimensional detector it is possible to separate the XRD (and thus the PDF) signal parallel and perpendicular to the applied electric field [97]. Different perovskites exhibit distinct behaviours: for $BaTiO_3$ and $SrTiO_3$, the electric field affects piezoelectric lattice strain and ionic polarizability, respectively, while $Na_{1/2}Bi_{1/2}TiO_3$ undergoes a sharp field-induced monoclinic to rhombohedral phase transition, accompanied by the reorientation of nanoscale ferroelectric domains.

Compared to other advanced techniques in solid-state chemistry, the PDF method has two main limitations. Firstly, obtaining high-quality PDF data—specifically with a high Q_{max} value and an adequate signal-to-noise ratio—requires synchrotron radiation. This dependence means that access to beam time is limited and not guaranteed, as each project undergoes evaluation, with beam time typically awarded months after the proposal is submitted. As an alternative to synchrotrons, several companies now offer laboratory diffractometers with Mo sources that can collect XRD data up to a Q_{max} of ~ 18 Å^{-1}. However, this Q_{max} is generally considered too low for detailed structural investigations, which means laboratory diffractometers are mostly used for screening or preliminary analysis. However, it should be noted that for fingerprinting analysis, especially with low-scattering compounds, collecting data up to very high Q_{max} values may be unnecessary or even detrimental. At high Q values, scattering factors approach zero, so extending the Q_{max} limit of the Fourier transform may increase noise rather than provide valuable structural insights [98]. The appropriate Q_{max} value should be determined for each specific case.

Another significant limitation of the PDF method is its lack of element sensitivity. While some advanced techniques enable the extraction of partial PDFs (which reflect only the contributions of a single element)—such as those combining X-ray and neutron diffraction [99], expensive isotopic substitution for neutron PDFs [100], or anomalous X-ray scattering [101]—PDF does not differentiate between elements. This becomes particularly important in the study of multicomponent or high-entropy materials. When structural information is needed about a specific element, other methods must be considered. A common alternative to PDF for investigating local structure is X-ray Absorption Spectroscopy (XAS). By analyzing the absorption coefficient of an element at its absorption edge (typically an L- or K-edge), XAS provides insights into the local structure of the element. Among other details, the edge and pre-edge signals (X-ray Absorption Near Edge Structure, or XANES) offer direct information about oxidation states and coordination, while the post-edge signal (Extended X-ray Absorption Fine Structure, or EXAFS) reveals details about radial distance distributions. Unlike PDF, EXAFS provides data on the local environment of specific species by selecting the suitable absorption edge. The advantage of XAS's element sensitivity is balanced by its focus on a limited structural range (typically only the first few coordination shells), making EXAFS a valuable complement to PDF. As an example, in the Gd-doped ceria solid solutions mentioned above, Ce-O and Gd-O distances are so close to each other that they appear as a single peak in the $G(r)$ curve. Conversely, EXAFS supplies two separated interatomic distances as well as the coordination number of each ion. Deguchi and coworkers demonstrated, collecting XAS spectra at the

Ce-K and Gd-K edges, that the Gd-O distances are ~0.05 Å shorter than Ce-O distances and oxygen vacancies are preferentially positioned close to the Gd ions [102]. However, like PDF, XAS also requires synchrotron radiation to gather high-quality data across a wide energy range. In view of their characteristics, they are complementary tools for describing structural disorder from the first neighbour to the nanometer scale, necessary when functional properties are controlled by complex defect architectures.

5 Disorder in Reciprocal Space: Single Crystal X-ray Diffraction

As anticipated in the previous section, if one is skilled (or lucky) enough to get a high X-ray quality single crystal, the disorder might be investigated directly in the reciprocal space by looking at the DS features. These manifest as nonzero continuous intensity lines, streaks or clouds in reciprocal lattice regions far from the Bragg points (Figs. 5 and 6) [103]. Such features come up with the TDS-derived ones, as well as the Bragg signals, resulting in total scattering signals that could be difficult to interpret properly.

Despite complicating the picture, disorder-induced DS is a further source of precious structural information for solid-state scientists. Anisotropic DS far from the Bragg reflections is often due to structured (or non-random) disorder (Fig. 5). For example, if defects are ordered in some way along certain crystallographic directions, or clustered in some real space planes, anisotropic DS signatures are produced, in the form of diffuse "lines" that interconnect the bright Bragg spots. These features relate to structural correlations at the nm-size scale [103], while in general DS maps the local departures of the molecular neighborhood from the average, long-range crystallographic structure.

Molecular crystals of high X-ray quality can be obtained with relatively lower efforts compared with metal oxides and ceramic materials; unfortunately, internal molecular degrees of freedom often complicate the interpretation of the observed features. Specific patterns of diffuse scattering can be correlated to the nature and extent of local disorder, including the preferred orientations of disordered groups, correlations between neighboring molecules, and potential energy barriers hindering molecular motion.

To the best of our knowledge, one of the first quantitative studies of DS features in organic crystals was pioneered by Welberry & Jones on 2,3-dichloro-6,7-dimethylanthracene in the late 1970s [106]. The next three decades witnessed the advent of position-sensitive detectors, which granted unprecedented resolution and accuracy of scattering intensities. More efficient and detailed computer models paralleled the experimental advances, allowing them to disentangle purely thermal DS from those due to structured disorder in complex materials.

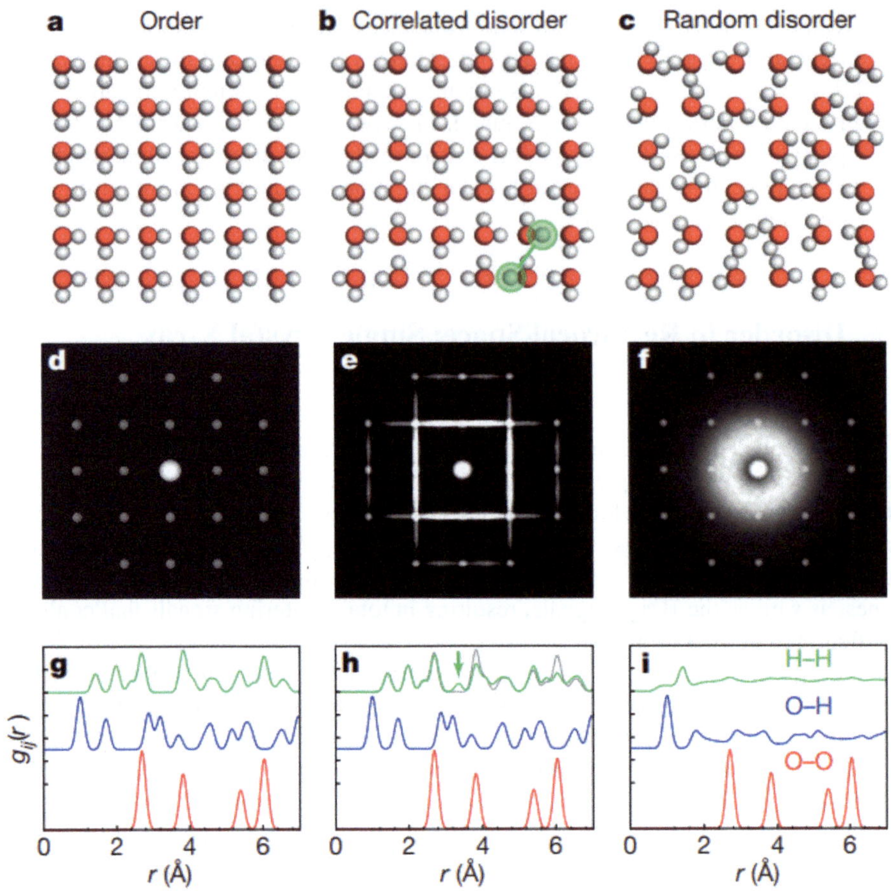

Fig. 5 Crystal packing for ordered (**a**) and disordered (**b**, **c**) water ice. (**d–f**) Show the corresponding Bragg and DS features in reciprocal space, while radial distribution functions for H-H, O-O and O-H contacts are shown as green, red and blue curves. (**g-i**) Reproduced with permission from [103]. (Courtesy of the original publisher ©)

For example, while diffuse scattering features in benzil ($C_{14}H_{10}O_2$) can be explained in terms of thermal librational motion of the molecular phenyl moiety, p-methyl-N-(p-chlorobenzylidene) aniline ($C_{14}H_{12}ClN$, ClMe) is highly disordered, with four competing basic orientations of the molecule in each crystallographic site [104]. DS was used also to study urea intercalation compounds (Fig. 6c), where long-chain alkanes like dibromodecane are accommodated into free hydrogen-bonded chains, forming a disordered 1-D pseudocrystal that produces characteristic DS stripes in the 0kl plane [105, 107].

The study of DS features is not restricted to the realm of X-rays. Recently, Gorelik and Schmidt [108] studied diffuse scattering from electron diffraction patterns recorded on a flexible organic pigment, orange 13 (PO13), through a 300 V

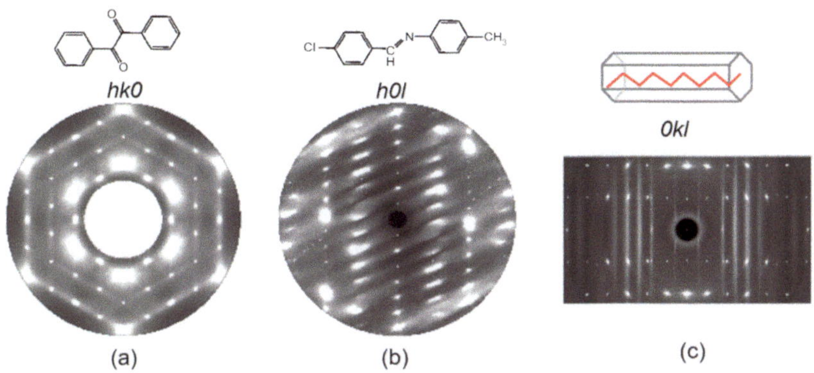

Fig. 6 (**a–b**) Observed XRD patterns in benzil (hk0 plane) and p-methyl-N-(p-chlorobenzylidene) aniline (h0l plane). Adapted with permission from [104]. (**c**) Vertical strikes in the 0kl plane of dibromodecane (DBD)/urea inclusion compound are due to disordered DBD molecules, which are allocated into urea channels and form a 1-D pseudo-crystal. Adapted with permission of the International Union of Crystallography from [105]. (Courtesy of the original publisher ©)

transmission electron microscope. The α phase of PO13 is usually commercialized as a nanocrystalline powder, which resisted previous attempts to solve the structure by standard crystallographic methods due to severe stacking disorder. Gorelik and Schmidt managed to solve the problem by looking at the whole scattering pattern, where DS features were explained by producing different structural models of stacked molecular layers with increasing thickness, and minimizing them by means of quantum methods.

6 Static or Dynamic Disorder? A Case Study

Structural disorder and atomic vibrations (see Sect. 2) can have similar effects in the description of equilibrium coordinates and in subtle features of residual density maps. According to current literature, an unambiguous consent on the actual interplay of cation disorder and anharmonicity of thermal motion in metal oxides does not exist, as they are strongly intertwined and cannot be easily disentangled through routine experimental techniques [109–111].

Pyrope ($Mg_3Al_2Si_3O_{12}$) is a recent case where static and dynamic disorder at the Mg site might produce analogue effects [21]. The problem was tackled through an experimental charge density (ρ) approach [112], relying on a very precise and accurate set of in-home measured single-crystal XRD amplitudes with a maximum resolution of 0.44 Å. The idea is that any statistically significant disorder of Mg^{2+} should be evident in accurate residual density Fourier maps, as both high-order and low order data carry information on possible deviations from the expected site stoichiometry.

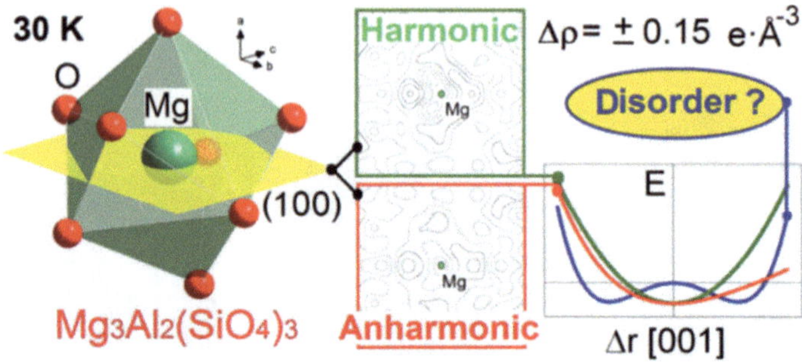

Fig. 7 Different views of the Mg^{2+} site in pyrope at 30 K. Left: coordination polyhedron (dodecahedron), center: residual 3 × 3 Å Fourier maps in the (100) plane after the multipole refinement without (top) and with (bottom) anharmonic contribution to the ADPs included. Contour levels are drawn at steps of 0.05 e $Å^{-3}$, with full (dotted) lines representing positive (negative) values. Right: example to illustrate the difference among a purely harmonic (parabolic green curve), an anharmonic (red curve) and a double-well (blue curve) potential. Reproduced with permission from [21]. (Courtesy of the original publisher ©)

In the mentioned work, no significant ($>2.5\,\sigma$) residual Fourier peaks were detectable from the experimental ρ around Mg, after least-squares refinement of a multipole model with anharmonic thermal motion at the Mg (Wyckoff 24c) site. The most relevant structured feature in the map (Fig. 7, center-bottom) consists of a pair of ρ maxima close to Mg, $+\,0.12$ e $Å^{-3}$ in magnitude and 0.58 Å apart from the ion, generating a 'shashlik-like' alternation of positive and negative ρ residuals.

What was at stake here was the interpretation of these structured residual charge density features: they could perhaps be associated with some kind of static positional disorder. However, any attempt to explicitly include 48f-disordered Mg in the multipole model, refining its fractional coordinates and monopole electronic parameters, invariably implied that the ion was brought back to its original, undisplaced 24c position within a few thousandths of Å. As a matter of fact, disorder—both static and dynamic—invariably implies that a double- (or multiple?)-well potential is present at the Mg 24c site (as shown in the right part of Fig. 7).

To unravel this point, the authors performed a full vibrational analysis of the pob-TZVP B3LYP normal modes in the *Ia3d* phase of perfect pyrope: the analysis showed that Mg is involved in several soft phonon modes, which imply large (>0.85 Å) relative stretching and bending displacements with respect to its surrounding O atoms.

Invoking the presence of some kind of dynamic disorder reconciles the intertwined views of anharmonicity and disorder. Such a model would indeed account for the highly anisotropic density features around the position 24c [21].

7 Tackling Disorder with Force Field Methods

A common feature of the works cited in Sects. 4 and 5 is the use of computational modeling techniques to complement experimental observations. Even though contemporary approaches take advantage of deep learning algorithms [113], classical Monte Carlo (MC) and Molecular Dynamics (MD) simulations remain the Queen approaches for understanding the complexity hidden in DS measurements data—and will likely continue to be so for a long time to come, due to the very large number of atoms or molecules needed to predict reliable DS features. This section does not intend to offer a comprehensive or detailed explanation of such techniques, as excellent references are available elsewhere [114, 115]. In a nutshell, MC and MD exploit equations and parameters, which allow one to compute the atom-atom potential energy in a pairwise fashion. Such parameters and equations constitute the so-called Force Field (FF). In this context, a "molecule" is a system made by incompressible spheres ("atoms") linked by flexible springs ("bonds"); long-range interactions are accounted for with standard potential forms, like the well-known Lennard-Jones and Coulombic terms. In MD, classical Newton or Lagrange equations are solved, positions and velocities of individual atoms are updated, and the whole system is developed through different "moves" to produce a "trajectory". In MC, no time-dependent equations are used. The system rather undergoes random moves of one degree of freedom at a time. Each move is rejected or accepted according to the so-called Metropolis criterion, where positive changes of total energy are discouraged but not completely hampered. This way, the system may climb kinetic barriers and explore all the relevant potential energy minima, eventually providing good estimates of equilibrium properties. If the FF parameters are accurate, there are reasonable chances that MC and MD trajectories disclose relevant structural and dynamic properties of the system.

FF-based tools have been also extensively employed in materials science for almost a century now. Among the first attempts, we recall the pioneering work by Mott and Littleton to simulate defects in solids, which dates to 1938 [116]. As the explicit modeling of the crystal lattice was impossible at that time, they defined a region of free relaxation in the neighboring of a defect, where a few lattice atoms are subject to explicit energy minimization. The defect region was embedded in a dielectric continuum, resulting in a reasonably good agreement of defect energies with experimental findings in ionic compounds. This achievement paved the way to the further simulation of a variety of equilibrium properties in ceramics and transition metal oxides, including defect distribution and clustering effects within the crystalline matrix [117–120]. The success of these early FF-based minimization algorithms is probably due to the dominance of strong ionic bonds in crystals where cohesion stems from the Coulomb attraction between oppositely charged particles. No significant competing forces exist, and FFs based on the Madelung potential are more than enough to accurately reproduce the experimental observations.

Of course, the main drawback is that energy minimizations give a static picture of the lattice at 0 K. Going hand in hand with the development of faster computers, in the 1980–1990s it has become more common to use MC or MD to explore the various accessible minima in the energy landscape of complex materials at finite temperature. Worth of note are applications to zeolite chemistry [121–123] and silica [124], for which phase changes and IR spectra were accurately predicted by classical simulations. A lot of efforts were also devoted to simulating the mass transport of various sorbates into zeolite channels, with various degrees of success in reproducing the experimental diffusion coefficients [125]. As expected, critical factors were dimensions and shape of the guest molecule, the pore diameter and the rigidity of the zeolite lattice.

In the last two decades, efforts have been made to model DS features from FF-based trajectories. MC is very effective in sampling different configurations of large defective systems, finding those that minimize the total energy and most probably contribute to disorder-induced DS. The coupling MC with neutron probes resulted very sensitive to small changes of the molecular packing, like those associated with ferroelectric phase transitions in triglycine sulfate [126]. Nowadays, MC is mature enough to treat also very large systems [71, 127] like the ZIF-90 molecular organic framework, where a combined X-ray/MC study showed that correlated disorder causes lattice distortions that affect both local symmetry and porosity [128]. However, MC may fail to account for DS features in more complex cases, for example whenever disorder and thermal motion are strongly intertwined, like in aspirin I [129]. In this case, it has been suggested to exploit MD simulations, as they can intrinsically account for correlated motions. Molecular Dynamics has indeed been used for a long time in the field of Materials Science [115], and it is still largely employed to study glasses and amorphous solids in general. These materials lack long-range order, making most experimental techniques blind to the atomic-level detail of their structure. Valuable information on the short-range organization comes from X-ray absorption (XAS) experiments and total scattering techniques, which give access to radial distribution functions (see above), and from solid-state spectroscopies (IR, NMR). MD can help to bridge the gap, for example by providing hints on the behavior of glass additives and partial ordering at the 10–100 nm scale. In silico models can be obtained by melting a crystalline precursor at high T, and then by rapidly quenching the fluid below the glass transition temperature.

MD simulations were able to recover the inner structure of vitreous SiO_2, as determined by neutron scattering experiments [130], and even self-diffusion rates of Si^{4+} and O^{2-} ions [131]. More complex glasses are formed when alkali metals are added as additives to the silica matrix. XAS studies show that these materials consist of alkali-rich clusters alternating with silica-rich regions [132]. If more than one alkali metal is present in the same matrix, different ions impede each other from hopping across available sites and the overall ionic conductivity of the glass is decreased [133]. Interestingly, both these structural and transport features were correctly reproduced by MD simulations based on FFs parametrized ad hoc [134]. More recently, the development of modern force fields, coupled with the exceptional increase of computing power, allowed the application of MD algorithms on the evaluation of intensive properties like specific heat [135] and anisotropic response properties [136], including non-equilibrium heat transfer [137], stress-strain effects [136, 138] and dynamic symmetry in molecular crystals [22, 23, 34].

Obviously, both techniques are as accurate as the FF they employ, as nicely summarized by the old programmers' motto GIGO ("garbage in, garbage out"). For sure, computational crystal chemistry has come a long way from estimating static cohesive energies in the 1970s, to the fine evaluation of complex static and thermal disorder in molecular crystals and disordered materials. The future probably holds even more fascinating challenges, for example the prediction of IR and Raman spectra of defective molecular materials, with potential applications in the accurate polymorph detection in pharmaceutical samples.

8 Conclusions

The importance of defects to understanding and rationalizing the observed properties of crystalline materials has been well known in the last century. However, it is only in the last two decades that the development of 4th generation synchrotron X-ray sources, paralleled by the even more rapid growth of hardware and software power, allowed materials scientists to really delve into the defect structure down to the atomic scale.

In this context, diffraction techniques still provide the most eligible methods to gain insights into nature and distribution of defects in the material. XRPD and total scattering experiments grant access to accurate information on the deviations between local and *long-range* crystal structure, providing scientists with very powerful tools to understand the subtleties of chemical coordination in solids. However, modern experimental approaches include also the interplay of calorimetric, spectroscopic and microscopic techniques, all of which being often supported by computer simulation methods. In fact, the focus is shifting beyond bare structural analysis, and it is increasingly oriented to understand how different defects are intertwined to each other in determining the behavior of the material. Ferroelectric relaxors and molecular crystals are instructive examples. In the former, the dynamics of interacting defects creates polar nanoregions that impact on the μm-length mesostructure and determine the phase diagram; in the latter, internal (molecular) degrees of freedom can be exploited to produce the desired movements, such as large librations of flat aromatic substituents or even intercalation structures.

Likely, the future will foresee the flourishing of even more advanced techniques, which will allow a finer control of the defect structure at all the length scales. The final goal is the top-down synthesis of tailored advanced materials, like—for example—solid-state molecular motors and reusable ceramic nanoscale memories.

Acknowledgments The Italian Ministry of Universities and Research (MUR) is acknowledged for partial funding this work through the PRIN—Research Projects of National Relevance initiative (project number 20223T577Z, title: FERROKNOTS—Harnessing ferroelectric topological defects for photonics).

References

1. R.J.D. Tilley, *Defects in Solids* (John Wiley & Sons, Hoboken, 2008)
2. A.M. Stoneham, *Theory of Defects in Solids: Electronic Structure of Defects in Insulators and Semiconductors* (Oxford University Press, Oxford, 2001)
3. N. Hannay, *Defects in Solids* (Springer Science & Business Media, New York, 2012)
4. M. Zhang, D. Yang, H. Luo, Q. Liang, J. Wu, X. Tang, Mater. Today Phys. **23**, 100645 (2022). https://doi.org/10.1016/j.mtphys.2022.100645
5. R. Borgoni, C. Galimberti, D. Zappa, Appl. Stoch. Models Bus. Ind. **37**, 878 (2021). https://doi.org/10.1002/asmb.2615
6. H.N. Wang, F. An, C.Y. Wong, K. Yin, J. Liu, Y. Wang, J.M. Zuo, A. Schleife, Q. Cao, Sci. Adv. **10** (2024). https://doi.org/10.1126/sciadv.adr8636
7. M. Kulichenko, N. Fedik, D. Steglenko, R.M. Minyaev, V.I. Minkin, A.I. Boldyrev, Chem. Phys. **532**, 110680 (2020). https://doi.org/10.1016/j.chemphys.2020.110680
8. H.L. Cui, G. Gumbs, Phys. Rev. B **42**, 7015 (1990). https://doi.org/10.1103/PhysRevB.42.7015
9. M.J. Cliffe, W. Wan, X. Zou, P.A. Chater, A.K. Kleppe, M.G. Tucker, H. Wilhelm, N.P. Funnell, F.X. Coudert, A.L. Goodwin, Nat. Commun. **5**, 4176 (2014). https://doi.org/10.1038/ncomms5176
10. D. Pierangeli, M. Ferraro, F.D. Mei, G.D. Domenico, C.E.M. de Oliveira, A.J. Agranat, E. DelRe, Nat. Commun. **7**, 10674 (2016). https://doi.org/10.1038/ncomms10674
11. L. Lo Presti, J. Parravicini, R. Soave, G. Parravicini, M. Mauri, L. Loconte, F. Di Mei, L. Falsi, L. Tartara, S. Binetti, et al., Phys. Rev. B **102**(21), 214110 (2020). https://doi.org/10.1103/PhysRevB.102.214110
12. F.C. Frank, W.T. Read, Phys. Rev. **79**, 722 (1950). https://doi.org/10.1103/PhysRev.79.722
13. P.M. Anderson, J.P. Hirth, J. Lothe, *Theory of Dislocations* (Cambridge University Press, Cambridge, 2017)
14. M. Kleman, J. Friedel, Rev. Mod. Phys. **80**, 61 (2008). https://doi.org/10.1103/RevModPhys.80.61
15. M.F. Chisholm, S. Kumar, P. Hazzledine, Science **307**, 701 (2005). https://doi.org/10.1126/science.1105962
16. J. McGervey, W. Triftshäuser, Phys. Lett. A **44**, 53 (1973). https://doi.org/10.1016/0375-9601(73)90957-2
17. L. Bai, D. Liu, X. Zhao, F. Yu, Y. Li, ACS Appl. Mater. Interfaces **15**, 3152 (2023). https://doi.org/10.1021/acsami.2c19480
18. T. Thonhauser, G.S. Jeon, G.D. Mahan, J.O. Sofo, Phys. Rev. B **68**(20), 205207 (2003). https://doi.org/10.1103/PhysRevB.68.205207
19. S.L. Semiatin, *Plastic Deformation Structures* (ASM International, Materials Park, 2005). https://doi.org/10.31399/asm.hb.v14a.a0004018
20. T.A. Read, Phys. Rev. **58**, 371 (1940). https://doi.org/10.1103/PhysRev.58.371
21. R. Destro, R. Ruffo, P. Roversi, R. Soave, L. Loconte, L. Lo Presti, Acta Crystallogr. Sect. B Struct. Sci. Cryst. Eng. Mater. **73**, 722 (2017). https://doi.org/10.1107/S2052520617006102
22. R. Destro, P. Roversi, M. Barzaghi, L. Lo Presti, Molecules **26**, 3075 (2021). https://doi.org/10.3390/molecules26113075
23. L. Sironi, G. Macetti, M. Vacchini, L. Lo Presti, Struct. Chem. (2025). https://doi.org/10.1007/s11224-025-02483-w
24. B.L. Hess, J.J. Ague, Geochim. Cosmochim. Acta **354**, 27 (2023). https://doi.org/10.1016/j.gca.2023.06.004
25. J.C. Mauro, *Diffusion in Crystals* (Elsevier, Amsterdam, 2021), pp. 129–146. https://doi.org/10.1016/B978-0-12-823907-0.00015-7
26. T. Ishigaki, S. Yamauchi, J. Mizusaki, K. Fueki, H. Naito, T. Adachi, J. Solid State Chem. **55**, 50 (1984). https://doi.org/10.1016/0022-4596(84)90246-9

27. R. Huang, Y. Wen, P. Miao, W. Shi, W. Niu, K. Sun, Y. Li, Y. Ji, B. Zhang, Chem. Catalysis **3**, 100667 (2023). https://doi.org/10.1016/j.checat.2023.100667
28. C.M. Zener, S. Siegel, J. Phys. Colloid Chem. **53**, 1468 (1949). https://doi.org/10.1021/j150474a017
29. A. Nowick, B. Berry, Acta Metall. **10**, 312 (1962). https://doi.org/10.1016/0001-6160(62)90008-1
30. C.P. Herrero, R. Ramírez, Chem. Phys. Lett. **568–569**, 70 (2013). https://doi.org/10.1016/j.cplett.2013.03.010
31. G. Macetti, L. Loconte, S. Rizzato, C. Gatti, L. Lo Presti, Cryst. Growth Des. **16**, 6043 (2016). https://doi.org/10.1021/acs.cgd.6b01069
32. O. Haida, T. Matsuo, H. Suga, S. Seki, J. Chem. Thermodyn. **6**, 815 (1974). https://doi.org/10.1016/0021-9614(74)90227-4
33. A. Gavezzotti, Acta Crystallogr. Sect. B Struct. Sci. Cryst. Eng. Mater. **78**, 333 (2022). https://doi.org/10.1107/S2052520621012191
34. A. Gavezzotti, L. Lo Presti, S. Rizzato, CrystEngComm **24**, 922 (2022). https://doi.org/10.1039/D1CE01360B
35. P.A. Fleury, J.F. Scott, J.M. Worlock, Phys. Rev. Lett. **21**, 16 (1968). https://doi.org/10.1103/PhysRevLett.21.16
36. N. Zhang, C. Gao, Y. Xiong, J. Energy Chem. **37**, 43 (2019). https://doi.org/10.1016/j.jechem.2018.09.010
37. M. Ceotto, L. Lo Presti, G. Cappelletti, D. Meroni, F. Spadavecchia, R. Zecca, M. Leoni, P. Scardi, C.L. Bianchi, S. Ardizzone, J. Phys. Chem. C **116**, 1764 (2012). https://doi.org/10.1021/jp2097636
38. L. Lo Presti, M. Ceotto, F. Spadavecchia, G. Cappelletti, D. Meroni, R.G. Acres, S. Ardizzone, J. Phys. Chem. C **118**, 4797 (2014). https://doi.org/10.1021/jp412394e
39. D. Meroni, C. Cionti, L. Silvestrini, N. Gal, M. Cazzaniga, M. Ceotto, G. Buccella, L. Lo Presti, G. Cappelletti, J. Photochem. Photobiol. A Chem. **444**, 114916 (2023). https://doi.org/10.1016/j.jphotochem.2023.114916
40. W.A. Plummer, AIP Conf. Proc. **17**, 147 (1974). https://doi.org/10.1063/1.2945918
41. W. Sprengel, B. Oberdorfer, E.M. Steyskal, R. Würschum, J. Mater. Sci. **47**, 7921 (2012). https://doi.org/10.1007/s10853-012-6460-9
42. Z. Shen, R. Wagoner, W. Clark, Acta Metall. **36**, 3231 (1988). https://doi.org/10.1016/0001-6160(88)90058-2
43. L. Lo Presti, V. Pifferi, G.D. Liberto, G. Cappelletti, L. Falciola, G. Cerrato, M. Ceotto, J. Mater. Chem. A **9**, 7782 (2021). https://doi.org/10.1039/D1TA01040A
44. Q. Liang, Z. Kloenne, Y. Zheng, D. Wang, S. Antonov, Y. Gao, Y. Hao, R. Yang, Y. Wang, H.L. Fraser, Scr. Mater. **177**, 181 (2020). https://doi.org/10.1016/j.scriptamat.2019.10.029
45. K. Zhussupbekov, L. Ansari, J.B. McManus, A. Zhussupbekova, I.V. Shvets, G.S. Duesberg, P.K. Hurley, F. Gity, C.O. Coileáin, N. McEvoy, npj 2D Mater. Appl. **5**, 14 (2021). https://doi.org/10.1038/s41699-020-00196-8
46. R. Würschum, B. Oberdorfer, E.M. Steyskal, W. Sprengel, W. Puff, P. Pikart, C. Hugenschmidt, R. Pippan, Phys. B Condens. Matter **407**, 2670 (2012). https://doi.org/10.1016/j.physb.2012.01.090
47. P. Scherrer, *Bestimmung der inneren Struktur und der Größe von Kolloidteilchen mittels Röntgenstrahlen* (Springer Berlin Heidelberg, Berlin, 1912), pp. 387–409. https://doi.org/10.1007/978-3-662-33915-2_7
48. G. Williamson, W. Hall, Acta Metall. **1**, 22 (1953). https://doi.org/10.1016/0001-6160(53)90006-6
49. A. Borbély, Scr. Mater. **217**, 114768 (2022). https://doi.org/10.1016/j.scriptamat.2022.114768
50. R.W. Cheary, A. Coelho, J. Appl. Crystallogr. **25**, 109 (1992). https://doi.org/10.1107/S0021889891010804
51. P. Scardi, Cryst. Growth Des. **20**, 6903 (2020). https://doi.org/10.1021/acs.cgd.0c00956
52. M. Wilkens, Phys. Status Solidi (a) **2**, 359 (1970). https://doi.org/10.1002/pssa.19700020224

53. M. Leoni, R.D. Maggio, S. Polizzi, P. Scardi, J. Am. Ceram. Soc. **87**, 1133 (2004). https://doi.org/10.1111/j.1551-2916.2004.01133.x
54. L. Balogh, G. Ribárik, T. Ungár, J. Appl. Phys. **100** (2006). https://doi.org/10.1063/1.2216195
55. Q.S. Paduano, D.W. Weyburne, A.J. Drehman, Phys. Status Solidi (a) **207**, 2446 (2010). https://doi.org/10.1002/pssa.201026258
56. H. Dittrich, M. Wohlfahrt-Mehrens, Int. J. Inorganic Mater. **3**, 1137 (2001). https://doi.org/10.1016/S1466-6049(01)00143-X
57. P. Scardi, M. Leoni, Acta Mater. **53**, 5229 (2005). https://doi.org/10.1016/j.actamat.2005.08.002
58. R.D. Shannon, Acta Crystallogr. Sect. A **32**, 751 (1976). https://doi.org/10.1107/S0567739476001551
59. N.Q. Minh, J. Am. Ceramic Soc. **76**, 563 (1993). https://doi.org/10.1111/j.1151-2916.1993.tb03645.x
60. R.B. Neder, F. Frey, H. Schulz, Acta Crystallogr. Sect. A Found. Crystallogr. **46**, 799 (1990). https://doi.org/10.1107/S0108767390006183
61. M. Coduri, M. Scavini, M. Allieta, M. Brunelli, C. Ferrero, Chem. Mater. **25**, 4278 (2013). https://doi.org/10.1021/cm402359d
62. D.N. Argyriou, J. Appl. Crystallogr. **27**, 155 (1994). https://doi.org/10.1107/S0021889893007964
63. M. Coduri, M. Scavini, M. Allieta, M. Brunelli, C. Ferrero, J. Phys. Conf. Ser. **340**, 012056 (2012). https://doi.org/10.1088/1742-6596/340/1/012056
64. M. Scavini, M. Coduri, M. Allieta, P. Masala, S. Cappelli, C. Oliva, M. Brunelli, F. Orsini, C. Ferrero, IUCrJ **2**, 511 (2015). https://doi.org/10.1107/S2052252515011641
65. R.W. Cheary, Acta Crystallogr. Sect. B Struct. Sci. **47**, 325 (1991). https://doi.org/10.1107/S0108768191000587
66. M. Coduri, A. Bernasconi, H.E. Fischer, L. Malavasi, J. Mater. Chem. A **8**, 21227 (2020). https://doi.org/10.1039/D0TA07073D
67. M.K. Crawford, M.A. Subramanian, R.L. Harlow, J.A. Fernandez-Baca, Z.R. Wang, D.C. Johnston, Phys. Rev. B **49**, 9198 (1994). https://doi.org/10.1103/PhysRevB.49.9198
68. R. Harlow, J. Res. Natl. Inst. Stand. Technol. **101**, 327 (1996). https://doi.org/10.6028/jres.101.034
69. C. Giacovazzo, *Fundamentals of Crystallography*, vol. 7 (Oxford University Press, Oxford, USA, 2002)
70. G.H. Stout, L.H. Jensen, *X-ray Structure Determination: A Practical Guide* (John Wiley & Sons, New York, 1989)
71. T. Welberry, T. Weber, Crystallogr. Rev. **22**, 2 (2016). https://doi.org/10.1080/0889311X.2015.1046853
72. U. Shmueli, *International Tables for Crystallography, Volume B: Reciprocal Space* (Springer Science & Business Media, New York, 2008)
73. B. Niepötter, R. Herbst-Irmer, D. Stalke, J. Appl. Crystallogr. **48**, 1485 (2015). https://doi.org/10.1107/S1600576715014934
74. S.J. Billinge, T. Egami, *Preface*, vol. 16, 2nd edn. (2012), pp. xi–xv. https://doi.org/10.1016/B978-0-08-097133-9.09990-1
75. S.J. Billinge, J. Solid State Chem. **181**, 1695 (2008). https://doi.org/10.1016/j.jssc.2008.06.046
76. R. Destro, L. Loconte, L. Lo Presti, P. Roversi, R. Soave, Acta Crystallogr. Sect. A Found. Crystallogr. **60**, 365 (2004). https://doi.org/10.1107/S0108767304014813
77. L. Lo Presti, D. Invernizzi, R. Soave, R. Destro, Chem. Phys. Lett. **416**, 28 (2005). https://doi.org/10.1016/j.cplett.2005.09.037
78. L. Lo Presti, M. Allieta, M. Scavini, P. Ghigna, L. Loconte, V. Scagnoli, M. Brunelli, Phys. Rev. B **84**, 104107 (2011). https://doi.org/10.1103/PhysRevB.84.104107
79. M.P. Anderson, D.E. Cox, K. Halperin, A.S. Nowick, Solid State Ion. **9–10**, 953 (1983). https://doi.org/10.1016/0167-2738(83)90116-9

80. F. Bertolotti, D. Moscheni, A. Guagliardi, N. Masciocchi, Eur. J. Inorganic Chem. **2018**, 3789 (2018). https://doi.org/10.1002/ejic.201800534
81. M. Scavini, F. Bertolotti, J. Mlloja, F. Umbri, A. Bosc, S. Cappelli, S. Checchia, C. Oliva, P. Fumagalli, D. Ceresoli, M. Longhi, A. Guagliardi, M. Coduri, Nanomaterials **12**, 3385 (2022). https://doi.org/10.3390/nano12193385
82. H.E. Fischer, A.C. Barnes, P.S. Salmon, Rep. Prog. Phys. **69**, 233 (2006). https://doi.org/10.1088/0034-4885/69/1/R05
83. P. Juhás, T. Davis, C.L. Farrow, S.J.L. Billinge, J. Appl. Crystallogr. **46**, 560 (2013). https://doi.org/10.1107/S0021889813005190
84. C.L. Farrow, P. Juhas, J.W. Liu, D. Bryndin, E.S. Božin, J. Bloch, T. Proffen, S.J.L. Billinge, J. Phys. Condens. Matter **19**, 335219 (2007). https://doi.org/10.1088/0953-8984/19/33/335219
85. P.J. Chupas, X. Qiu, J.C. Hanson, P.L. Lee, C.P. Grey, S.J.L. Billinge, J. Appl. Crystallogr. **36**, 1342 (2003). https://doi.org/10.1107/S0021889803017564
86. B.H. Toby, T. Egami, Acta Crystallogr. Sect. A Found. Crystallogr. **48**, 336 (1992). https://doi.org/10.1107/S0108767391011327
87. G. Confalonieri, V. Buscaglia, G. Canu, M.T. Buscaglia, M. Dapiaggi, J. Synchrotr. Radiat. **26**, 1280 (2019). https://doi.org/10.1107/S1600577519004508
88. X. Qiu, E.S. Božin, P. Juhas, T. Proffen, S.J.L. Billinge, J. Appl. Crystallogr. **37**, 110 (2004). https://doi.org/10.1107/S0021889803026670
89. S.J.L. Billinge, *Real-Space Rietveld: Full Profile Structural Refinement of the Atomic Pair Distribution Function* (Kluwer Academic Publishers, Boston, 2002), pp. 137–156. https://doi.org/10.1007/0-306-47077-2_8
90. X. Qiu, T. Proffen, J.F. Mitchell, S.J.L. Billinge, Phys. Rev. Lett. **94**, 177203 (2005). https://doi.org/10.1103/PhysRevLett.94.177203
91. M. Coduri, P. Masala, M. Allieta, I. Peral, M. Brunelli, C.A. Biffi, M. Scavini, Inorganic Chem. **57**, 879 (2018). https://doi.org/10.1021/acs.inorgchem.7b02896
92. G.H. Kwei, A.C. Lawson, S.J.L. Billinge, S.W. Cheong, J. Phys. Chem. **97**, 2368 (1993). https://doi.org/10.1021/j100112a043
93. V. Petkov, M. Gateshki, M. Niederberger, Y. Ren, Chem. Mater. **18**, 814 (2006). https://doi.org/10.1021/cm052145g
94. A.A. Bokov, Z.G. Ye, J. Mater. Sci. **41**, 31 (2006). https://doi.org/10.1007/s10853-005-5915-7
95. C.W. Ahn, C.H. Hong, B.Y. Choi, H.P. Kim, H.S. Han, Y. Hwang, W. Jo, K. Wang, J.F. Li, J.S. Lee, I.W. Kim, J. Korean Phys. Soc. **68**, 1481 (2016). https://doi.org/10.3938/jkps.68.1481
96. I.K. Jeong, J.K. Lee, Appl. Phys. Lett. **88**, 262905 (2006). https://doi.org/10.1063/1.2217162
97. T.M. Usher, I. Levin, J.E. Daniels, J.L. Jones, Sci. Rep. **5**, 14678 (2015). https://doi.org/10.1038/srep14678
98. T. Dykhne, R. Taylor, A. Florence, S.J.L. Billinge, Pharm. Res. **28**, 1041 (2011). https://doi.org/10.1007/s11095-010-0350-0
99. S. Sampath, C.J. Benmore, K.M. Lantzky, J. Neuefeind, K. Leinenweber, D.L. Price, J.L. Yarger, Phys. Rev. Lett. **90**, 115502 (2003). https://doi.org/10.1103/PhysRevLett.90.115502
100. J. Bréger, N. Dupré, P.J. Chupas, P.L. Lee, T. Proffen, J.B. Parise, C.P. Grey, J. Am. Chem. Soc. **127**, 7529 (2005). https://doi.org/10.1021/ja050697u
101. V. Petkov, I.K. Jeong, F. Mohiuddin-Jacobs, T. Proffen, S.J.L. Billinge, W. Dmowski, J. Appl. Phys. **88**, 665 (2000). https://doi.org/10.1063/1.373718
102. H. Deguchi, H. Yoshida, T. Inagaki, M. Horiuchi, Solid State Ion. **176**, 1817 (2005). https://doi.org/10.1016/j.ssi.2005.04.043
103. D.A. Keen, A.L. Goodwin, Nature **521**, 303 (2015). https://doi.org/10.1038/nature14453
104. T.R. Welberry, Mol. Cryst. Liquid Cryst. **440**, 1 (2005). https://doi.org/10.1080/15421400590957576
105. T.R. Welberry, Acta Crystallogr. Sect. B Struct. Sci. Cryst. Eng. Mater. **78**, 344 (2022). https://doi.org/10.1107/S2052520621012865

106. T.R. Welberry, R.D.G. Jones, J. Appl. Crystallogr. **13**, 244 (1980). https://doi.org/10.1107/S0021889880012009
107. T.R. Welberry, *Diffuse X-ray Scattering and Models of Disorder*, vol. 31 (Oxford University Press, Oxford, 2022)
108. T.E. Gorelik, S.L. Bekö, J. Teteruk, W. Heyse, M.U. Schmidt, Acta Crystallogr. Sect. B Struct. Sci. Cryst. Eng. Mater. **79**, 122 (2023). https://doi.org/10.1107/S2052520623000720
109. C.A. Geiger, Am. Mineral. **98**, 780 (2013). https://doi.org/10.2138/am.2013.4301
110. A. Nakatsuka, M. Shimokawa, N. Nakayama, O. Ohtaka, H. Arima, M. Okube, A. Yoshiasa, Am. Mineral. **98**, 783 (2013). https://doi.org/10.2138/am.2013.4402
111. M.T. Dove, Am. Mineral. **82**, 213 (1997). https://doi.org/10.2138/am-1997-3-401
112. R. Destro, L. Lo Presti, R. Soave, A.E. Goeta, *Multi-temperature Electron Density Studies* (Springer Netherlands, Cham, 2011), pp. 659–696. https://doi.org/10.1007/978-90-481-3836-4_19
113. C.A. Fuller, L.S.P. Rudden, IUCrJ **11**, 34 (2024). https://doi.org/10.1107/S2052252523009521
114. W.F. van Gunsteren, H.J.C. Berendsen, Angew. Chem. Int. Ed. Engl. **29**, 992 (1990). https://doi.org/10.1002/anie.199009921
115. G. Macetti, L. Sironi, L. Lo Presti, *Classical Molecular Dynamics Simulation of Molecular Crystals and Materials: Old Lessons and New Perspectives* (Elsevier, Cham, 2024), pp. 777–803. https://doi.org/10.1016/B978-0-12-821978-2.00107-0
116. N.F. Mott, M.J. Littleton, Trans. Faraday Soc. **34**, 485 (1938). https://doi.org/10.1039/tf9383400485
117. C.R.A. Catlow, B.E.F. Fender, J. Phys. C Solid State Phys. **8**, 3267 (1975)
118. C.R.A. Catlow, A.N. Cormack, F. Théobald, Acta Crystallogr. Sect. B Struct. Sci. **40**, 195 (1984). https://doi.org/10.1107/S010876818400197X
119. S.C. Parker, C.R.A. Catlow, A.N. Cormack, Acta Crystallogr. Sect. B Struct. Sci. **40**, 200 (1984). https://doi.org/10.1107/S0108768184001981
120. C.R.A. Catlow, *Defects in Solids: Modern Techniques* (Springer, New York, 1986), pp. 269–301
121. S. Tsuneyuki, Y. Matsui, H. Aoki, M. Tsukada, Nature **339**, 209 (1989). https://doi.org/10.1038/339209a0
122. M.K. Song, J.M. Shin, H. Chon, M.S. Jhon, J. Phys. Chem. **93**, 6463 (1989). https://doi.org/10.1021/j100354a037
123. J.B. Nicholas, A.J. Hopfinger, F.R. Trouw, L.E. Iton, J. Am. Chem. Soc. **113**, 4792 (1991). https://doi.org/10.1021/ja00013a012
124. J.S. Tse, D.D. Klug, J. Chem. Phys. **95**, 9176 (1991). https://doi.org/10.1063/1.461198
125. P. Demontis, G.B. Suffritti, Chem. Rev. **97**, 2845 (1997). https://doi.org/10.1021/cr950253o
126. J.M. Hudspeth, D.J. Goossens, T.R. Welberry, M.J. Gutmann, J. Mater. Sci. **48**, 6605 (2013). https://doi.org/10.1007/s10853-013-7457-8
127. D.J. Goossens, M.J. Gutmann, Phys. Rev. Lett. **102**, 015505 (2009). https://doi.org/10.1103/PhysRevLett.102.015505
128. S. Canossa, R. Frison, X. Pei, H.B. Bürgi, Acta Crystallogr. Sect. A Found. Adv. **77**, C1001 (2021). https://doi.org/10.1107/S0108767321087006
129. E.J. Chan, J. Appl. Crystallogr. **48**, 1420 (2015). https://doi.org/10.1107/S1600576715013242
130. B. Vessal, J. Non-Crystall. Solids **177**, 103 (1994). https://doi.org/10.1016/0022-3093(94)90520-7
131. D.A. Litton, S.H. Garofalini, J. Non-Crystall. Solids **217**, 250 (1997). https://doi.org/10.1016/S0022-3093(97)00107-5
132. G.N. Greaves, Philos. Mag. B **60**, 793 (1989). https://doi.org/10.1080/13642818908209743
133. G. Greaves, Solid State Ion. **105**, 243 (1998). https://doi.org/10.1016/S0167-2738(97)00471-2
134. B. Vessal, G.N. Greaves, P.T. Marten, A.V. Chadwick, R. Mole, S. Houde-Walter, Nature **356**, 504 (1992). https://doi.org/10.1038/356504a0

135. S. Engelmann, R. Hentschke, Sci. Rep. **9**, 7606 (2019). https://doi.org/10.1038/s41598-019-44132-3
136. S. Rizzato, A. Gavezzotti, L. Lo Presti, Cryst. Growth Des. **20**, 7421 (2020). https://doi.org/10.1021/acs.cgd.0c01098
137. X. Chen, A. Munjiza, K. Zhang, D. Wen, J. Phys. Chem. C **118**, 1285 (2014). https://doi.org/10.1021/jp410054j
138. X. Du, H. Zhao, L. Zhang, Y. Yang, H. Xu, H. Fu, L. Li, Sci. Rep. **5**, 16275 (2015). https://doi.org/10.1038/srep16275

Meaning and Applications of Fröhlich Entropy

Jacopo Parravicini, Eugenio DelRe, and GianBattista Parravicini

Abstract We discuss the rich physical meaning of Fröhlich entropy. Namely, we discuss how it can be used to pinpoint three different features of a given dielectric, i.e. the dipolar orientation disorder, the global state-of-order, and the level of stability of a given phase. Specific validity hypotheses and practical uses are also considered.

1 Basic Concepts and Assumptions

A formal definition of Fröhlich entropy (FE) was already given in chapters "Introduction: Dielectric Calorimetry and Fröhlich Entropy Estimation" and "Elements of Thermodynamics of Dielectrics and Basic Theory of Fröhlich Entropy". We here start by briefly recalling that, under the previously discussed conditions (Sect. 3.1 in chapter "Introduction: Dielectric Calorimetry and Fröhlich Entropy Estimation"), the entropy S of a dielectric material subject to an electric field E may be written as the sum of two terms, i.e.[1]

[1] Where S_0 is the field independent entropy variation and, as usual, T is the temperature, ε is the absolute dielectric function, which is the product of the ε_0 vacuum dielectric constant and ε_s, the relative dielectric function.

J. Parravicini (✉)
Dipartimento di Fisica & Astronomia, Università degli Studi di Firenze, Sesto F., Firenze, Italy
e-mail: jacopo.parravicini@unifi.it

E. DelRe
Dipartimento di Fisica, "Sapienza" Università di Roma, Roma, Italy
e-mail: eugenio.delre@uniroma1.it

GB. Parravicini
Dipartimento di Fisica, Università degli Studi di Pavia, Pavia, Italy
e-mail: gianbattista.parravicini@unipv.it

J. Parravicini (ed.), *Fröhlich Entropy Estimation and Dielectric Calorimetry*,
Advances in Dielectrics, https://doi.org/10.1007/978-3-032-21454-6_5

$$S(T, E) = S_0(T) + S_E(T, E) = S_0 + \frac{E^2}{2}\frac{\partial \varepsilon}{\partial T}, \tag{1}$$

so FE is

$$S_E = \frac{E^2}{2}\frac{\partial \varepsilon}{\partial T}, \tag{2}$$

and, for unitary field,

$$s(T) = \frac{1}{2}\frac{\partial \varepsilon}{\partial T}. \tag{3}$$

In short, we may affirm that, when the hypotheses of Eq. (1) are satisfied, *FE is the variation of the entropy of an (ideal) dielectric under the application of a (static) electric field* [1, 2]. In order to suitably discuss the physical meaning of the FE concept, we recall here the phrases by Fröhlich himself [1].[2]

> Finally, equation 12 [$S = S_0(T) + (\partial\varepsilon_s/\partial T)(E^2/8\pi)$, i.e. Eq. (1) above] shows that the entropy is increased by the field if $\partial\varepsilon_s/\partial T$ is positive, and decreased if this quantity is negative. Since the entropy is a measure of the molecular disorder, an external field creates order in dipolar liquids and gases for which ε_s decreases with increasing T. This may be expected because the field will orientate some of *the dipoles which in the absence of a field are at random*. In some dipolar solids, on the other hand ε_s increases with T, which means that an external field increases the disorder. *This too is understandable if one assumes that in the absence of a field the dipoles are in a well-ordered state as may be expected in solids.* The field by turning some of the dipoles into different directions can thus only decrease the existing order.[3]

We note Fröhlich's interpretation was recovered in other monographs on dielectrics by other important scholars, namely Scaife and by Böttcher, that expressed the same concepts with very similar words[4] [3–5]. Fröhlich illustrates these statements through a schematic drawing analogous to our plot in Fig. 1, which the author describes with the following words [1].

> Schematic temperature-dependence of the dielectric constant ε_s and of the entropy change $S \propto \partial\varepsilon_s/\partial T$ due to the polarization by a field. If $S > 0$ the field creates disorder, if $S < 0$ it creates order. Near the absolute zero of temperature the substance is *already* perfectly ordered. Hence $\partial\varepsilon_s/\partial T$ cannot be negative near $T = 0$.[5]

Hence, the first interpretation of FE is given by Fröhlich himself. Actually, such a picture assumes entropy S as a statistical measure of disorder. The higher the disorder (or the lower the degree of order), the larger is the entropy. While classical equilibrium thermodynamics defines the entropy as

---

[2] Chapter "Introduction: Dielectric Calorimetry and Fröhlich Entropy Estimation", pp. 12–13.

[3] Emphasis added.

[4] See, in particular, chapter 3 § 13 of [3].

[5] Emphasis added.

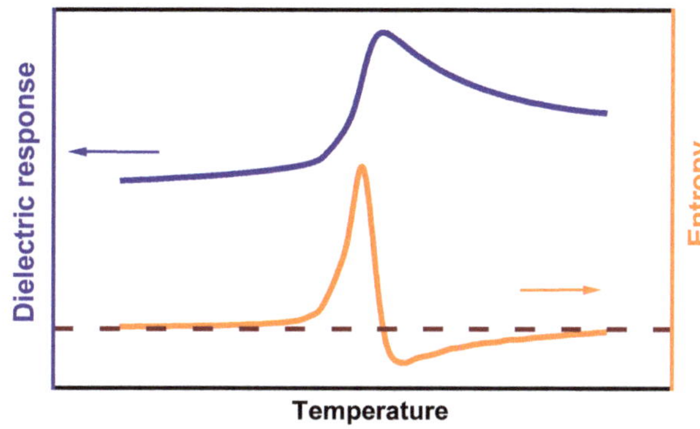

Fig. 1 Ideal scheme of the dielectric constant versus temperature (violet curve) and of the entropy variation calculated from Eq. (2) (orange curve). Dotted horizontal line signals the 0 value of FE

$$dS = \frac{\delta Q}{T} = C_p \frac{dT}{T}, \tag{4}$$

where δQ is the amount of heat absorbed by the considered physical system in a reversible process, and C_p is the specific heat (at constant pressure). We know that statistical mechanics theory is able to provide the entropy of a system as a function of its internal states, so that

$$S = k_B \log w, \tag{5}$$

where w is the number of microstates of the physical system. In this sense, the standard basic interpretation of entropy correlates disorder with the number of ways in which the occupation of the states inside a system can be microscopically arranged, where each different configuration can be considered a microstate of a thermodynamic state [6, 7].

On the other hand, we have to remember that a straightforward interpretation of the entropy as a "figure of merit of disorder" is not considered satisfactory, since it assumes different meanings for different conditions and investigated systems [8, 9]. Nevertheless, several experiments repeatedly confirmed that such a quantity, in the contexts considered by Fröhlich, provides a suitable picture of the state of the dielectric following the above-recalled plain Fröhlich's interpretation.

If we follow the temperature evolution of the field-induced entropy, we may obtain information about the system "in the absence of field". This concept is the basis for the experimental application of FE measurement. Actually, the analysis of the FE trend versus temperature can highlight the unperturbed state of the investigated dielectric material, provided that E is a low-strength *probe field*. This interpretation points out that the state-of-order of a dielectric system can be described by exploiting the behavior of its (static) relative permittivity on temperature. When field-induced entropy is estimated, it gives the conceptual basis to develop a form of dielectric calorimetry and thermal analysis, which has been called *Fröhlich entropy evaluation* [2].

What is the exact physical meaning of *Fröhlich entropy*? Here it is worth considering that this concept has a quite general validity: FE has been used to investigate and characterize very different physical systems, such as metallic nanoparticles [10–12], nematic compounds [13–18], molecular crystals [19, 20], ferroelectric disordered perovskites [21–25], polar glasses [26], polar and glass-forming liquids [27–31], polymeric films [32], enzymes [33]. In these studies, at least three specific physical meanings of FE can be identified.

2 Physical Meaning of Fröhlich Entropy

The experimental use highlights different aspects of FE. These interpretations are very general, so they can be applied to a large class of compounds; of course, the fulfilling of the basic hypotheses of Fröhlich's theory (listed and discussed in Sect. 3.1 in chapter "Introduction: Dielectric Calorimetry and Fröhlich Entropy Estimation") is the necessary condition allowing to use it [2]. Actually, the assumption of thermodynamic equilibrium implies static conditions, so no time-dependent fields in principle can be considered [27, 34]); this requires that only the real part of the dielectric function $\mathrm{Re}(\varepsilon)$ is relevant. In fact, if the imaginary part of the dielectric response $\mathrm{Im}(\varepsilon)$ were not negligible, relevant relaxation phenomena would occur [35–38], so temporal variables and correlated dynamics would be strongly involved. This would invalidate Fröhlich's reasoning and his consequent interpretation. In real systems $\mathrm{Im}(\varepsilon)$ is never null, so, in practical terms, we need to compare the real and imaginary parts of the dielectric function. This can be done through a physical quantity that is usually considered in dielectric spectroscopy, i.e. the loss tangent,

$$\tan\delta = \frac{\mathrm{Im}(\varepsilon)}{\mathrm{Re}(\varepsilon)}, \tag{6}$$

which was previously recalled in Sect. 3.2 in chapter "Introduction: Dielectric Calorimetry and Fröhlich Entropy Estimation". The experimental conditions for the validity of FE interpretations is the same as the applicability of Fröhlich's theory (Eq. (11)), i.e. [7]

$$|\tan\delta| \ll 1. \tag{7}$$

Moreover, the strength of E should also be evaluated, and it should be maintained, at a level of a mere *probe* field, far from coercive values [7, 27, 38, 39].

The three physical meanings of FE which were identified in experiments are the following. The *first* meaning is associated to the simplest conceptual framework for a dielectric, i.e. a set of dipoles which are oriented by the application of an electric field. The *second* interpretation is focused on the description of phases and phase transitions, while the *third* one is correlated to the evaluation of the stability degree of a considered phase. As hinted above, all these perspectives have been practically taken into account during the experimental investigation of specific physical systems, where the FE has been calculated and discussed.

2.1 Fröhlich Entropy as a Figure-of-Merit of Orientation Disorder

The equations which contain internal energy, entropy, and Helmholtz free energy discussed in the previous chapter[6] have important and profound physical meanings. Here, the most important aspects are correlated to the entropy provided by Eq. (2).[7] This relationship points out that applying the electric field increases the entropy of the dielectric when $\partial\varepsilon_s/\partial T > 0$, while the same field causes a decrease of entropy when the derivative is negative ($\partial\varepsilon_s/\partial T < 0$). Let's focus on Fröhlich's original statements recalled above.

The simplest model of a dielectric material is a set of randomly-oriented electric dipoles which are free to move, in position or orientation (or both); this is the case for most of liquids and gases, which can be considered to contain (and display) a low content of structural information [9]. The dielectric permittivity of similar disordered systems typically decreases when the temperature increases [27]. Within Fröhlich's framework, however, the temperature dependence of the dielectric permittivity is not interpreted as a consequence of the degree of order; rather, its temperature derivative is used as a diagnostic criterion to distinguish liquid-like and solid-like dipolar states through the sign of the field-induced entropy variation. In such conditions the applied external field compels the dipoles to collectively align, so introducing regularities and increasing the information content: following Fröhlich, we say that it *creates order* [1]. A second, somewhat opposite, large class of dielectrics is represented by physical systems that are composed of dipoles in a well ordered state; these can have more or less fixed positions. Such compounds are typically found in crystalline solids, whose components are in stable minima of energy. In such conditions the application of an external field perturbs a stable pattern by orienting some dipoles into different directions: in Fröhlich's words, the action of the electric field enhances the amount of disorder, which is signaled by the increase of ε_s on T.

Both these ideal behaviors are shown in the scheme in Fig. 1. All these considerations show that the calculation of FE provides a measurement of the degree-of-freedom of orientation, i.e. it is an *orientation entropy*, which is able to point out the degree of rearrangement of the dipoles in the system, treated here as effective descriptors of the underlying charge distribution [29, 40, 41]. It is worth noting that this conceptual framework is not limited to systems with effective dipoles

[6] Chapter "Introduction: Dielectric Calorimetry and Fröhlich Entropy Estimation".

[7] Or even Eq. (3).

(e.g. polar liquids [27, 29]). On the contrary, it can be exploited to describe all compounds whose internal charge distribution can be reasonably approximated in terms of dipoles.

2.2 *Fröhlich Entropy as a Suitable Indicator of the State-of-Order*

A second interesting interpretation of Fröhlich entropy is the detailed description on temperature of the state-of-order of the physical system. As above recalled this can be done when time relaxation phenomena are negligible.[8] A virtue of this figure of merit is the possibility of a continuous temperature detection. Actually, the fact that FE is extrapolated from dielectric measurements as a function of temperature with an arbitrary rate allows us to follow continuously the trend and changes of the order state. This highlights the features of the phases and the occurring of possible transitions. FE evaluation is in this manner particularly suitable to detect and analyze phase transitions along with many of their features. If we investigate the transition region versus temperature, the $\varepsilon = \varepsilon(T)$ function can display either a kink or, in most cases, a characteristic peak [39]; the width of such structures provides an estimation of the temperature extension of the process and of its features. Usually transitions display variation either of the order state or order amount.

When FE is defined, an order/order transition is identified by a change of the positive slope of the $s = s(T)$ trend. Following Fröhlich's rationale, this is due to the fact that the field produces an entropy contribution which is positive both before and after the transition with different amounts. Moreover, the correlated numerical values of $s(T)$ can be interpreted as *a degree of order* of each phase [1, 4, 10, 19, 20]. Analogously, when a disorder/disorder transition occurs, $\varepsilon(T)$ and $s(T)$ show similar slopes, with negative values of FE.

Conversely, when an order/disorder (or disorder/order) transition arises, $s(T) \propto \partial\varepsilon_s/\partial T$ curve usually exhibits two-opposed-peaks, as in the ideal pattern shown in Fig. 1. This shape corresponds to positive and negative (or negative and positive) slopes in the related dielectric permittivity, whose negative part highlights a disordered state in the considered physical system. A typical example of such behavior is shown, e.g., in Fig. 2, in the temperature region around 217 K. These features can be found in a very large class of systems, making FE estimation a powerful figure of merit to evaluate the change of order degree in large class of physical systems in condensed matter.

[8] i.e. under the previously exposed Fröhlich's hypotheses on equilibrium and not-time-dependent fields.

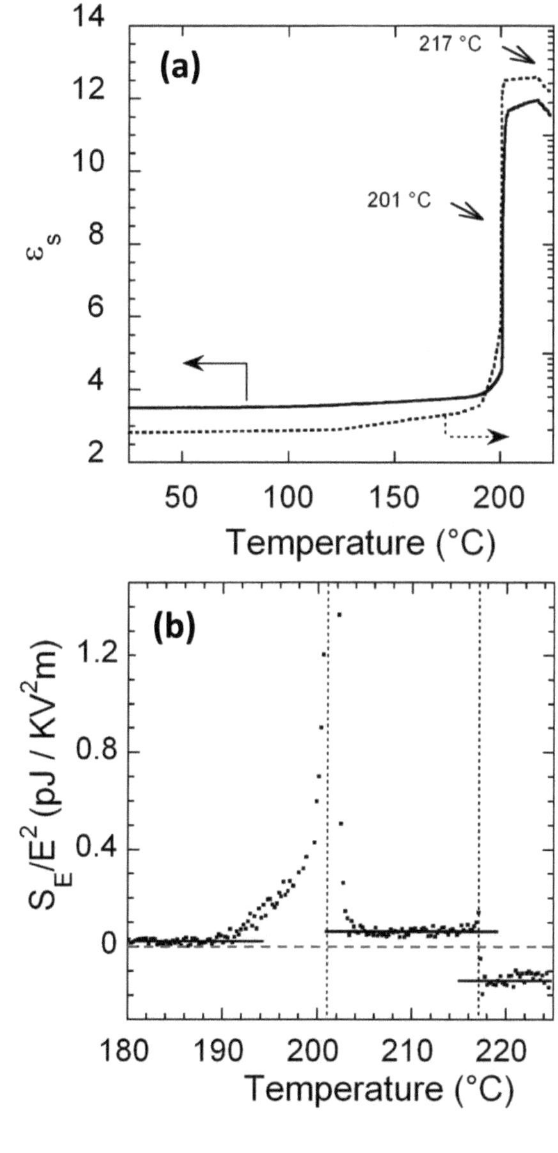

Fig. 2 (**a**) measured dielectric response ε_s and ac conductivity σ (in logarithmic scale) as a function of temperature T for an α-quaterthiophene single crystal. (**b**) FE contribution S_E/E^2 as a function of temperature T, obtained from the derivative of ε_s in the temperature region 180–225 °C, where a solid-solid and a solid-liquid transitions arise; the horizontal lines indicate averaged derivative values before and after the two main transitions, while the vertical dotted lines indicate the transition temperatures (adapted from [19], Fig. 2, courtesy of the original publisher ©)

2.3 *Fröhlich Entropy as a Quantitative Evaluator of Phase Stability*

A third interpretation of the physical meaning of FE is the quantitative evaluation of the stability degree of a phase as a function of temperature. Such an approach is still based on the Fröhlich's interpretation and it was especially proposed and discussed in experimental works on organic molecular crystals [19, 20].

When the electric field is applied to the sample, if this causes a large amount of entropy variation, $|s(T)|$, this will point out that the system is greatly modified by such an external perturbation. Conversely, if, in the same conditions, the obtained $|s(T)|$ is relatively modest, this is due to the fact that the external perturbation slightly modifies the energetic pattern of the system. If we put together these simple observations, we may reasonably deduce that in the last case the system lies in a more stable configuration than in the first one. Therefore, we conclude that, both for ordered ($s > 0$) and disordered ($s < 0$) conditions, FE can give a direct quantitative assessment of the stability degree of a given phase. Such stability can be described as a function of temperature and can be compared between different phases. In this vein, it is worth noting that an approximately constant $s = s(T)$ signals a range of temperature where the investigated system is stable; on the other hand, a steep variation of $s(T)$ can be associated to instability features. Such features usually become marked close to the phase transition regions, so they can be assumed as fingerprints of pre- and post- transitional regimes.

In order to better grasp this conceptual approach, we may focus on the ideal behavior depicted in Fig. 1. The plot shows that the largest values of FE are found in proximity of the phase transition, where the physical system has a strongly unstable global state due to the changing of its configuration. The farther we go away from the transition region, the more the phase becomes stable. This implies that the electric field causes an almost negligible perturbation to the system [38]: this is pointed out by the small relative amount of FE, which is decreasing and changing slower and slower while a stable global phase is established.

Another interesting example can be found in the FE response from measured permittivity of a disordered inorganic perovskite, displayed in Fig. 3 [42]; Fig. 4, on the other hand, shows the response of an organic molecular crystal [20]. In both cases, signs of phase instability are provided by the presence of $s(T)$ oscillations on temperature. Conversely, the data displayed by Fig. 2, which refer to an organic semiconductor, highlight a regular (nearly constant) behavior of FE, which is paradigmatic of highly stable phases [19]. However, the presence of oscillations of FE points out that the inherent dipoles of the considered physical system are reconfiguring. It is worth noting that when such oscillations cross the $S_E = 0$ value, they signal the establishing of mixed and metastable phases where order and disorder coexist. In these conditions, the order degree of the system is evolving, rapidly fluctuating from disordered to ordered state (Fig. 3).

3 Experimental Use of Fröhlich Entropy Estimation

3.1 Landscape

In previous chapters (especially in chapter "Experimental Techniques to Investigate the State-of-Order in Condensed Matter") we saw that many experimental

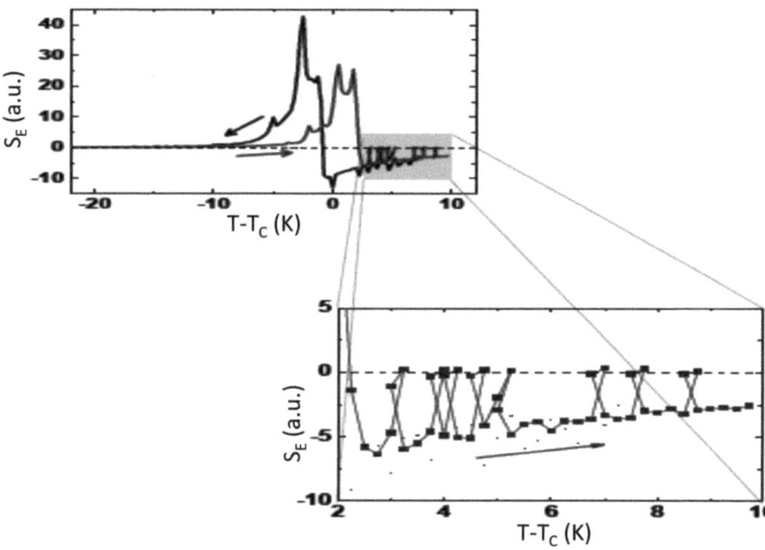

Fig. 3 (**a**) Fröhlich entropy (here indicated with S_E, per volume unit, for a unitary electric field) close to the Curie point of a disordered potassium-based inorganic perovskite crystal and (**b**) zoom of shaded region showing phase fluctuations from disordered-like to ordered-like in the heating stage (adapted from Supplementary Information of [42], Fig. 2)

approaches may be used to investigate the degree of order/disorder of a material. Which approach should be used depends on a number of factors, including the class of systems being evaluated, the required sensitivity, the temperature and/or temporal region to be investigated, the size and costs of the equipment, etc. In general, standard known techniques can be classified on the basis of the characteristic spatial scale of the information they provide, i.e. microscopic (e.g. X-ray diffraction), mesoscopic (e.g. TEM), and macroscopic-averaged (e.g. spectroscopy) ones (e.g. see the discussion in chapter "Experimental Techniques to Investigate the State-of-Order in Condensed Matter").

Morphology and order state are generally investigated through techniques such as X-ray diffractometry [43–45] (Sects. 1, 2, 4, and 5 in chapter "Experimental Techniques to Investigate the State-of-Order in Condensed Matter"), EXAFS [46, 47] (Sect. 2.1 in chapter "Fröhlich Entropy of Nanoparticles: Investigation of Phases, Phase Transitions and Melting"), X-ray absorption spectroscopy [48], neutron diffractometry [49, 50], electronic diffraction [51, 52], and nuclear magnetic resonance [45, 53, 54]. When these methods are employed in temperature-resolved mode, they can describe the inherent symmetry arrangement of the sample and the correlated evolution (and/or dynamics) when the macroscopic thermodynamic parameters are altered. Since these approaches are primarily based on the study of reciprocal space, their speciality is providing information on periodicity, where the existence of a certain classes of crystalline structures—or lack them—can be examined.

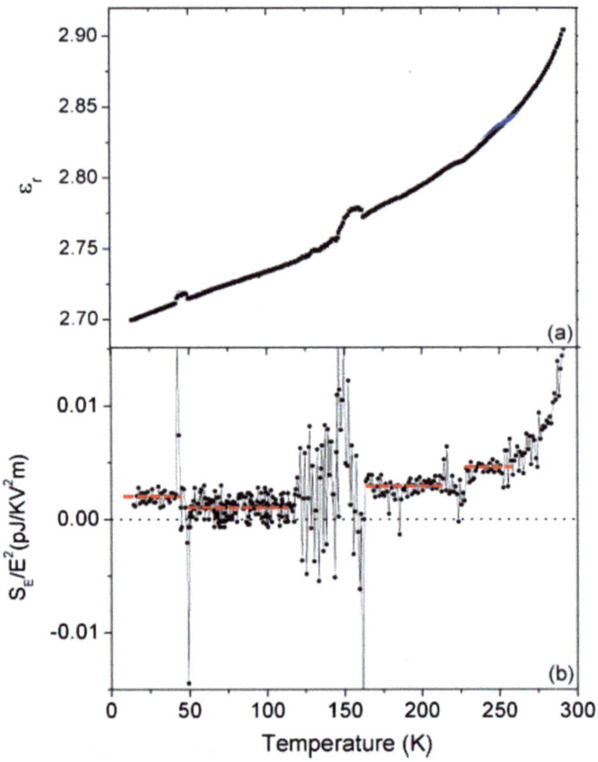

Fig. 4 (**a**) dielectric constant ε_r on T of a 3-μm-thick tetracene single crystal. (**b**) $s = s(T)$, obtained from the derivative of the dielectric constant of plot (**a**) (adapted from [20], Fig. 1, courtesy of the original publisher ©)

On the other hand, thermal analysis and calorimetric methods, which globally measure the macroscopic energy levels, are based on different principles [55–57]. Among the most important physical phenomena where both the temperature and the concept of state-of-order play fundamental roles, are phase transitions. These raise a large interest, because they involve several physical concepts, such as stability, critical temperatures, thresholds, amplitudes, structural symmetries, anisotropies, etc. Calorimetry and thermal analysis techniques[9] are widely employed to investigate phase transitions in condensed matter, while dielectric analysis and spectroscopy are typically used to test electric and transport features [7, 39, 58]. Likewise, dielectric measurements are also suitable to provide information similar to those obtained by calorimetric techniques: in fact, the relationships correlating dielectric and thermodynamic variables can be exploited to obtain information about the state, state-of-order, and evolution of a given investigated sample [2].

[9] See e.g. Table 2.

3.2 Remarks

Summarizing, the technique of FE estimation can be seen as a reliable experimental option to investigate several physical systems in condensed matter. In this line of thinking, the thermodynamic relationships which correlate the electromagnetic physical quantities, notably electric and magnetic fields, susceptivity and susceptibility, dielectric displacement and magnetization fields, stresses and strains with temperature and thermodynamic potentials [59], are the starting point to develop the theory and suitable interpretations which can be practically exploited. The relations first developed by Fröhlich [1, 60] and subsequently generalized [61], giving the possibility of calculating the entropy induced by the application of a static electric field, are the most important results from a practical point of view, providing a powerful tool. However, this requires the basic Fröhlich's interpretation (Sect. 1 pg. 1) to be accurately grasped. The above discussed interpretations of FE allow to observe several features, such as:

1. almost continuous thermal evolution of the state-of-order of a sample;
2. the level of stability of an explored phase;
3. the details of phase transitions, critical temperatures, and ordering features;
4. information about correlation state and rotational degree-of-freedom of dipoles inside a materials;
5. directional investigation highlighting anisotropies (see chapter "Generalized Fröhlich Entropy: Anisotropic and Nonlinear Media").

Therefore, FE physical quantity can be applied in condensed matter as a relatively straightforward way to investigate a large amount of different physical systems. Exploiting FE represents a useful practical approach complementary to the other experimental methods.

References

1. H. Fröhlich, *Theory of Dielectrics*, 2nd edn. (Claredon Press, Oxford, UK, 1958)
2. J. Parravicini, G. Parravicini, Res. Phys. **28**, 104571 (2021)
3. C.J. Boettcher, *Theory of Electric Polarization. Dielectrics in Static Fields* (Elsevier, Amsterdam, 1973)
4. B.K.P. Scaife, *Principles of Dielectrics*, 2nd edn. (Claredon Press, Oxford, UK, 1998)
5. B.K.P. Scaife, *Principles of Dielectrics*, 1st edn. (Claredon Press, Oxford, UK, 1989)
6. L. Brillouin, *Science and Information Theory* (Academic Press, New York, 1962)
7. K. Kao, *Dielectric Phenomena in Solids - With Emphasis on Physical Concepts of Electronic Processes* (Elsevier Academic Press, San Diego, 2004)
8. K. Denbigh, Br. J. Philos. Sci. **40**, 323 (1989)
9. A. Ben Naim, *A Farewell to Entropy - Statistical Thermodynamics Based on Information*, 1st edn. (World Scientific, Singapore, 2008)
10. GB. Parravicini, A. Stella, P. Tognini, P. Merli, A. Migliori, P. Cheyssac, R. Kofman, Appl. Phys. Lett. **82**, 1461 (2003)

11. GB. Parravicini, A. Stella, M. Ungureanu, P. Merli, A. Migliori, P. Cheyssac, R. Kofman, Eur. Phys. J. D **24**, 219 (2003)
12. GB. Parravicini, A. Stella, M. Ungureanu, P. Merli, A. Migliori, R. Kofman, Phys. Status Solidi B **237**, 374 (2003)
13. J. Jadzỳn, G. Czechowski, J. Phys. Chem. B **111**, 3727 (2007)
14. J. Jadzỳn, U. Sokołowska, J. Déjardin, J. Phys. Chem. B **112**, 9050 (2008)
15. J. Jadzỳn, G. Czechowski, J. Déjardin, J. Phys. Chem. B **112**, 4948 (2008)
16. J. Jadżyn, J. Swiergiel, D. Bauman, R. Dąbrowski, K. Czupryńskic, U. Sokołowska, Phase Transit. **82**, 409 (2009)
17. D. Bauman, J. Jadżyn, E. Wolarz, A. Modlińska, R. Dąbrowski, Opto-Electron. Rev. **18**, 63 (2010)
18. J. Świergiel, I. Płowaś́, J. Jadżyn, J. Mol. Liq. **220**, 879 (2016)
19. A. Sassella, D. Braga, M. Campione, T. Ciabattoni, M. Moret, J. Parravicini, GB. Parravicini, J. Appl. Phys. **109**, 013529 (2011)
20. G.B. Parravicini, M. Campione, F. Marabelli, M. Moret, A. Sassella, Phys. Rev. B **86**, 024107 (2012)
21. J. Parravicini, E. DelRe, A. Agranat, GB. Parravicini, Phys. Rev. B **93**, 094203 (2016)
22. P. Tan, H. Tian, F. Huang, X. Meng, Y. Wang, C. Hu, X. Cao, L. Li, Z. Zhou, Phys. Rev. Appl. **11**, 024037 (2019)
23. L. Lo Presti, J. Parravicini, R. Soave, GB. Parravicini, M. Mauri, L. Loconte, F.D. Mei, L. Falsi, L. Tartara, S. Binetti, A.J. Agranat, E. DelRe, Phys. Rev. B **102**, 214110 (2020)
24. J. Parravicini, E. DelRe, A.J. Agranat, GB. Parravicini, Nanoscale **9**, 9572 (2017)
25. F. Huang, C. Hu, H. Tian, X. Meng, P. Tan, Z. Zhou, CrystEngComm **21**, 7002 (2019)
26. D.V. Matyushov, J. Chem. Phys. **145**, 034504 (2016)
27. G.P. Johari, J. Chem. Phys. **138**, 154503 (2013)
28. G.P. Johari, J. Chem. Phys. **139**, 026101 (2013)
29. G.P. Johari, J. Chem. Phys. **145**, 164502 (2016)
30. S. Samanta, R. Richert, J. Chem. Phys. **142**, 044504 (2015)
31. S. Samanta, R. Richert, J. Phys. Chem. B **120**, 7737 (2016)
32. V. Kochervinskii, S. Bedin, I. Razumovskaya, M. Shcherbina, M. Buzin, I. Malyshkina, A. Prazdnichnyi, A. Pavlov, O. Vasilevskii, Polym. Sci. Ser. A **58**, 345 (2016)
33. Y. Kurzweil-Segev, A. Greenbaum (Gutina), I. Popov, D. Golodnitsky, Y. Feldman, Phys. Chem. Chem. Phys. **18**, 10992 (2016)
34. R. Richert, J. Phys. Condens. Matter **29**, 363001 (2017)
35. W.T. Coffey, Y.P. Kalmykov (eds.), *Fractals, Diffusion, and Relaxation in Disordered Complex Systems: Advances in Chemical Physics, Part B* (John Wiley & Sons, Hoboken, 2006)
36. S.J. Rzoska, V.P. Zhelezny (eds.), *Nonlinear Dielectric Phenomena in Complex Liquids* (Kluwer Academic Publisher, New York, 2004)
37. P. Lunkenheimer, M. Koehler, S. Kastner, A. Loidl, *Dielectric Spectroscopy of Glassy Dynamics* (John Wiley & Sons, Hoboken, 2012), chap. 3, pp. 115–149. https://doi.org/10.1002/9781118202470.ch3
38. P. Lunkenheimer, F. Humann, D. Reuter, K. Geirhos, A. Loidl, G. Johari, Phys. Rev. Mater. **3**, 112601 (2019)
39. F. Jona, G. Shirane, *Ferroelectric Crystals* (Dover, New York, 1993)
40. J.P. Sethna, *Statistical Mechanics: Entropy, Order Parameters, and Complexity*, 1st edn. (Oxford University Press, Oxford, UK, 2006)
41. J. Świergiel, J. Jadżyn, Phys. Chem. Chem. Phys. **13**, 3911 (2011)
42. E. DelRe, F. Di Mei, J. Parravicini, GB. Parravicini, A.J. Agranat, C. Conti, Nat. Photonics **9**, 228 (2015)
43. R. Comès, M. Lambert, A. Guinier, Solid State Commun. **6**, 715 (1968)
44. R. Comès, M. Lambert, A. Guinier, Acta Crystallogr. Sect. A **26**, 244 (1970)
45. K. Hirota, S. Wakimoto, D. Cox, J. Phys. Soc. Jpn. **75**, 111006 (2006)
46. G. Faraci, in *Encyclopedia of Nanoscience and Nanotechnology*, ed. by H. Nalwa (American Scientific Publishers, Valencia, 2004)

47. J. Stöhr, *NEXAFS Spectroscopy* (Springer, Berlin, 1991)
48. G. Buncker, *Introduction to XAFS* (Cambridge University Press, Cambridge, 2010)
49. D. Orauttapong, J. Toulouse, J. Robertson, Z. Ye, Phys. Rev. B **64**, 212101 (2001)
50. G. Xu, G. Shirane, J. Copley, P. Gehring, Phys. Rev. B **69**, 064112 (2004)
51. K. Tsuda, R. Sano, M. Tanaka, Phys. Rev. B **86**, 214106 (2012)
52. K. Tsuda, R. Sano, M. Tanaka, Appl. Phys. Lett. **102**, 051913 (2013)
53. B. Zalar, V. Laguta, R. Blinc, Phys. Rev. Lett. **90**, 037601 (2003)
54. B. Zalar, A. Lebar, J. Seliger, R. Blinc, V. Laguta, M. Itoh, Phys. Rev. B **71**, 064107 (2005)
55. P. Haines, *Principles of Thermal Analysis and Calorimetry* (Royal Society of Chemistry, Cambridge, UK, 2002)
56. M. Brown, *Introduction to Thermal Analysis*, 2nd edn. (Kluwer Academic Press, New York, 2001)
57. X. Yan, P. Liaw, Y. Zhang, Metall. Mater. Trans. A **52**, 2111 (2021)
58. M. Lines, A. Glass, *Principles and Applications of Ferroelectrics and Related Materials* (Oxford University Press, Oxford, UK, 2001)
59. J.F. Nye, *Physical Properties of Crystals - Their Representation by Tensors and Matrices* (Clarendon Press, Oxford, 1985)
60. H. Fröhlich, *Theory of Dielectrics*, 1st edn. (Clarendon Press, Oxford, UK, 1949)
61. J. Parravicini, Physica B **541**, 54 (2018)

Fröhlich Entropy of Nanoparticles: Investigation of Phases, Phase Transitions and Melting

GianBattista Parravicini, Paolo Ghigna, and Jacopo Parravicini

Abstract We discuss the application of Fröhlich Entropy estimation as a dielectric thermodynamic tool to investigate phases, phase transitions, and melting processes in low-dimensional systems. In the first section we summarize selected results obtained on Gallium nanoparticles by means of an established powerful technique such as X-ray Absorption Spectroscopy (XAS). Firstly, we sketch the different phenomena that are associated with X-ray absorption by matter, and outline the theory of XAS. Then, we illustrate how the nanometric structure drives the phase transitions of Ga particles, as a function of their size. In particular, we will show how small nanoparticles (5 nm in radius) can be supercooled in the liquid state down to 90 K, and how larger nanoparticles (10 and 30 nm in radius) crystallize in the high pressure δ-Ga phase. These results are interpreted in terms of surface tension. Secondly, in the following section, Fröhlich Entropy estimation is discussed in relation to temperature-resolved dielectric measurements, extending the analysis beyond metallic nanoparticles. Specifically, this approach is shown to be useful for highlighting phase transitions, pretransitional phenomena, premelting, and melting processes, providing detailed information on the evolution of the state of order and phase stability, in excellent agreement with the data obtained by the XAS technique.

GB. Parravicini
Dipartimento di Fisica, Università degli Studi di Pavia, Pavia, Italy
e-mail: gianbattista.parravicini@unipv.it

P. Ghigna
Dipartimento di Chimica, Università degli Studi di Pavia, Pavia, Italy
e-mail: paolo.ghigna@unipv.it

J. Parravicini (✉)
Dipartimento di Fisica & Astronomia, Università degli Studi di Firenze, Sesto F., Firenze, Italy
e-mail: jacopo.parravicini@unifi.it

J. Parravicini (ed.), *Fröhlich Entropy Estimation and Dielectric Calorimetry*,
Advances in Dielectrics, https://doi.org/10.1007/978-3-032-21454-6_6

1 Introductory Concepts

1.1 General View

Nanoparticles are at the center of extensive researche due to their current and foreseen use in nanotechnology. Structural characterization of nanoparticles is a prominent issue, as nanometric structures can trigger the formation of unstable or metastable phases. This is the case, for example, of TiO_2, whose surface free energy tends to stabilize the anatase structure rather than the equilibrium rutile structure [1]. Several probes can be used to gain information on the structure of nanosized materials. These can be grouped into:

1. diffraction techniques: e.g. X-ray diffraction, electron diffraction, etc.;
2. microscopic techniques: e.g. HRTEM, STEM, etc.
3. spectroscopic techniques: e.g. optical spectroscopy, Raman spectroscopy, dielectric spectroscopy, etc.

In this respect, *X-ray absorption spectroscopy* (XAS) has several advantages, making it an ideal structural probe for nanoparticles:

- it is element selective, allowing to study complex chemical compositions element by element;
- it is very sensitive, thus requiring a minimal amount of sample and operating at very low concentrations;
- it can be performed on every state of matter, which is very useful when nanoparticles are not crystalline.

On the other hand, temperature-resolved dielectric spectroscopy is intrinsically sensitive to surface phenomena, so it can be easily adapted to investigate different kind of samples, and it is relatively simple and not expensive to be implemented.

1.2 The Peculiar Properties of Metallic Ga

Gallium is one of the four non-radioactive elements that are found to be liquid at temperatures close to room temperature, the others being caesium, rubidium, and mercury. For this property, and for its low reactivity, metallic Ga is widely used as a safer substitute for mercury, e.g. as a thermometric fluid: the melting point of Ga, (302.9146 K) is used as a reference point in the International Temperature Scale [2]. The crystal structure of solid Ga is also unique, as it does not crystallize in any of the simple crystal structures. On the contrary, Ga has an orthorhombic crystal structure (space group *Abma*(3)) with 8 atoms in the unit cell. The structure is highly anisotropic and each atom has just one nearest neighbor 2.44 Å apart, the bonding between neighbors being covalent. The structure thus encompasses Ga_2 dimers, and this atypical crystal structure is responsible for the fact that the molar volume of the solid is larger than that of the liquid, a property that gallium shares with few other substances, including water, silicon, germanium and bismuth. It should be noted, moreover, that several polymorphs are found at high pressures [3].

Liquid Ga has a marked tendency to undercool below the melting point, and this property can be enhanced by the nanostructural configuration. Below we review the main findings on the undercooling of Ga nanoparticles, as a result of studies which combine diffractometric and spectroscopic techniques. After that, we will summarize some other results on the influence of nanostructure on Ga polymorphism.

1.3 The Ga Nanoparticles

When Ga is considered in nanometric-sized forms, it is often studied in a configuration where nanoparticles are embedded in a SiO_x matrix. This arrangement is considered paradigmatic, analogously to other confined systems where interface phenomena play key roles in determining their peculiar properties [4, 5]. Moreover, the study of the changes in phase transitions in correlation with the nanometric status of this metal is an interesting topic, that has been faced in several works both from a theoretical and experimental point of view [4–7]. In this vein, the rich polymorphism exhibited by Ga makes it one of the most interesting elements in condensed-matter domain. Namely, several works establish that nanometric Ga particles display several distinct structural phases, which can be more or less stable. Apart from the standard crystalline one, that is called Ga α-phase, these peculiar phases are usually named β, γ, δ, and ϵ [4, 5, 7–9]. Among these works, the experimental ones that employ spectroscopic techniques show that the dielectric investigation is able to distinguish the various solid phases of Ga nanoparticles in a confined environment [6, 10]. Therefore, the investigation of such physical systems can be accomplished by combining the above-recalled techniques, in order to explore both the microscopic and macroscopic domains, by diffractive and microscopy methods on one side, and spectroscopic methods on the other side.

2 Structural Characterization of Nanoparticles with Synchrotron Radiation Techniques

In this section, we first deal with the foundations of XAS, and then we demonstrate how the potential of XAS can be fully exploited in the study of Ga nanoparticles.

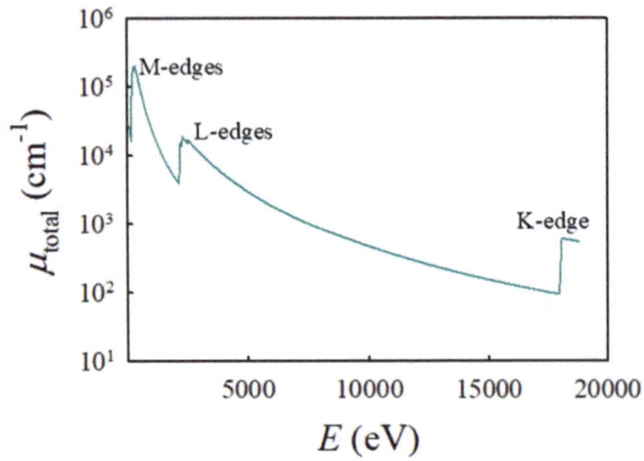

Fig. 1 X-ray absorption coefficient for Zirconium on a wide energy scale

2.1 XAS Fundamentals

In XAS, an X-ray photon is absorbed. The energy of X-rays matches that of atomic core levels, and therefore the absorption involves core electrons. The absorption is reflected in sharp edges in the absorption coefficient when plotted vs. the incoming photon energy on a large energy scale (see Fig. 1). The origin of the edge is straightforward: the edge energy position matches that of an atomic core level. If the photon energy is lower than this threshold energy, no absorption is detected. For higher photon energies, the absorption causes photoionization from a core level and the expulsion of a core electron. The edges are named according to the principal quantum number n of the core level involved. Thus, photoionization from $n = 1$ levels correspond to the K edge, from $n = 2$ levels correspond to the L edges, from $n = 3$ levels correspond to the M edges, etc. For $n \geq 2$, considering the atomic final state, it is readily understood that multiple edges at different energies arise. We work out here the simplest case of $n = 2$, but generalization is straightforward. Photoionization from the 2s orbital leads to the 2s^1 electronic configuration, corresponding to one state and to the L_I edge. Photoionization from the $2p$ orbitals leads to the $2p^5$ electronic configuration, corresponding to six states. Spin-orbit coupling with $L = 1$ and $S = 1/2$ gives $J = 1/2$ and $3/2$, corresponding to the L_{II} and L_{III} edges, accounting for 2 and 4 states, respectively. Finally, considering orbital energy and Hund's rules, the edge energies are ordered as $L_I > L_{II} > L_{III}$. This description is sufficient for explaining the X-ray absorption spectra on a large energy scale such as that of Fig. 1.

On a finer energy scale, the spectrum shows finer structures that are usually treated as separate spectroscopies. In particular, the energy region from −20 to +50, 100 eV around the rising edge is called *X-ray absorption near edge structure* (XANES) or *near-edge X-ray absorption fine structure* (NEXAFS), while the region extending to larger energies is called *extended X-ray absorption fine structure*(EXAFS). This partition, although in some way conventional, reflects the fact that different phenomena are at the basis of XANES and EXAFS. In this section, we first consider EXAFS as its explanation is simpler.

In the EXAFS regime, the photon energy is well above the photoionization threshold, and therefore the kinetic energy of the photoelectron is quite large. The photoelectron can be seen as a spherical wave exiting the photoabsorber, with wavevector **k**:

$$|\mathbf{k}| = k = \sqrt{\frac{2m_e}{\hbar^2}(\mathcal{E} - \mathcal{E}_0)}, \tag{1}$$

where m_e is the electron mass, $\mathcal{E}$ is the actual energy of the X-ray photon, and $\mathcal{E}_0$ is the edge energy. If the spherical wave e^{ikr}/kr hits a neighboring atom at a distance R, it can be backscattered with amplitude $f(k)$ and phaseshifts $\delta(k)$. The two waves can interfere, and it can be proved that constructive (destructive) interference leads to maxima (minima) in the absorption coefficient. Thus, we can define the EXAFS as:

$$\chi(k) = \frac{e^{ikR}}{kR} 2kf(k)e^{i\delta(k)} \frac{e^{ikR}}{kR} + \text{c.c.} \tag{2}$$

Accounting for a Gaussian disorder in the distance R_j with variance σ_j, considering the photoelectron mean free path for inelastic scattering $\lambda(k)$, and summing over the coordination shells with coordination number N_j, one finally gets

$$\chi(k) = \sum_j \frac{1}{kR_j} \left[N_j f_j(k) e^{-2R_j/\lambda(k)} e^{-2k^2\sigma_j^2}\right] \sin\left[2kR_j + \delta_j(k)\right]. \tag{3}$$

This is the EXAFS equation that can be used for fitting the experimental data, and obtaining the shell parameters R_j, σ_j and N_j, once the scattering amplitudes $f_j(k)$ and phaseshifts $\delta_j(k)$ are known. The EXAFS interpretation is then straightforward, as it is readily seen that the modulus of the EXAFS Fourier transform (FT) has a meaning similar to that of a radial distribution function. The difference lies in the fact that the peak amplitude in the EXAFS FT is influenced also by the scattering amplitudes $f(k)$ in addition to N_j, while the peak position is shifted with respect to R_j by $\delta(k)$, which usually is of the order of 0.5 Å.

The interpretation of XANES is less simple, as different phenomena overlap in this energy region. One is the spectroscopic transition to empty (valence) states from the core orbital. In this respect, XANES is an invaluable tool for probing the density of empty valence states as projected onto the photoabsorber. This projection arises because the transition dipole moment integral $\langle\psi_f|\boldsymbol{\mu}|\psi_i\rangle$ contains the initial state core level ψ_i wavefunction, and this is sufficiently large in the close proximity of the photoabsorber only.

The other phenomenon is due to multiple scattering of the photoelectron wave, which has a large amplitude for small $|\mathbf{k}|$. Usually, only three- and four- legged scattering paths are significant, but this means that XANES can be used for retrieving information about high (three and four body) correlation functions that

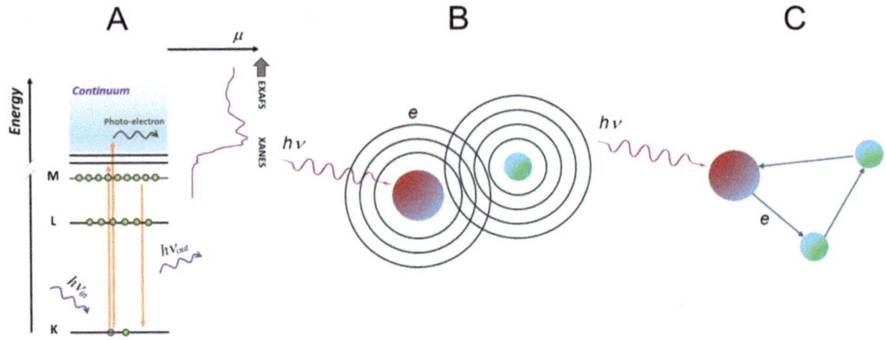

Fig. 2 Schematic picture of the phenomena of X-ray absorption. (**a**) Absorption of an X-ray photon from core levels, and emission of X-ray fluorescence. (**b**) single scattering of the photoelectron in the EXAFS regime. (**c**) a three legged multiple scattering path in the XANES regime

are hardly obtainable with other experiments, especially in disordered systems such as liquids or glasses. All the considerations explained in this section are depicted in a schematic way in Fig. 2.

2.2 *Experimental*

Details on the preparation of the nanoparticles and films have been reported elsewhere [11]. TEM and electron diffraction measurements were performed on samples cooled by liquid nitrogen in a FEI Tecnai F20 instruments. The XAS experiments were performed in fluorescence mode at the GILDA beamline (*European Synchrotron Radiation Facility*, ESRF, Grenoble) at the Ga-K edge. A Si(311) double crystal monochromator was used; harmonic rejection was realized by Pd mirrors, having a cutoff energy of 20 keV. A 13-elements high purity germanium detector was used for the collection of the fluorescence spectra, with the samples at 45 ° with respect to the incoming beam. The measurements were performed as a function of temperature in the range from 90 to 350 K after cooling the four films down to 88 K. The EXAFS data analysis was performed by means of the EXCURVE program [12].

2.3 *On the Undercooling of Metallic Ga Nanoparticles*

The undercooling of Ga nanoparticles is investigated by means of XANES at the Ga K-edge. As stated above, XANES is a probe for three and four body correlation functions, and therefore it is a valuable tool for investigating phase transitions in

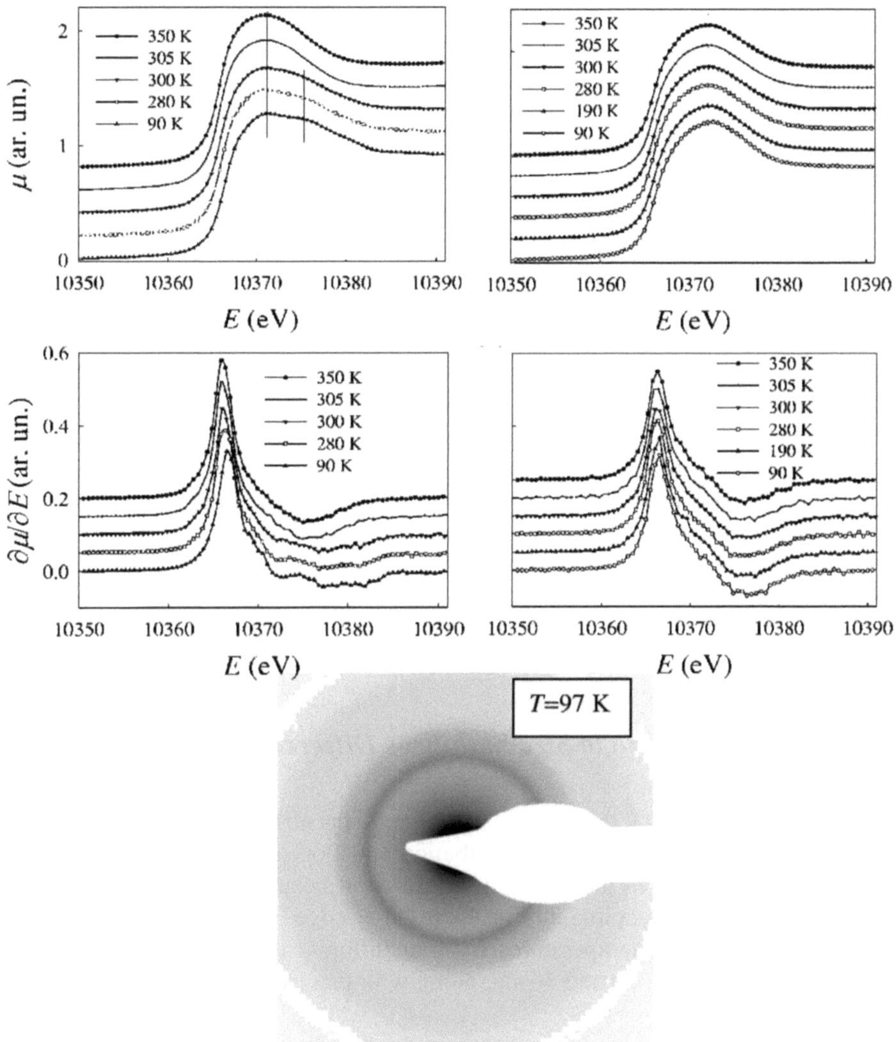

Fig. 3 Top-left: Ga K edge XANES and the corresponding derivatives of the continuous film at different temperatures as indicated. Top-right: same but for the 5 nm nanoparticles. Bottom: Electron diffraction pattern of the sample of the 5 nm sample at T = 97 K. Adapted with permission from [9]. (Courtesy of the original publisher © American Institute of Physics)

matter [9]. The experiments are carried out at 350, 305, 300, 280, 190 and 90 K, on two samples: a continuous film, ca. 300 nm thick, used as a reference for bulk Ga, and a thin film consisting of isolated nanoparticles with a radius of ca. 5 nm. The XANES results are shown in Fig. 3. The difference between the continuous film and the nanoparticles is striking. The continuous film shows a marked change in shape going from 350 K, which is well above the melting point, to 90 K: a marked shoulder at ca. 10375 eV appears below 305 K, indicating the transition from the liquid to the solid phase. On the contrary, the XANES spectrum for the 5 nm nanoparticles displays an unchanging spectral shape with changing temperature. In addition, this spectral shape is identical to that of the continuous film at 350 K, thus pointing to the presence of a liquid phase in the whole temperature range. This conclusion is supported by the electron diffraction pattern taken at 97 K, also shown in Fig. 3, where no well-defined diffraction effects are visible, and displaying only the diffuse scattering typical of a disordered phase.

The existence of a liquid gallium phase at 90 K points to an undercooling $\Delta T/T_M \geq 0.7$ which has been unprecedented before. This extreme value for the undercooling can be explained in two ways. First, we note that nano-droplets with a radius of 5 nm have a large surface tension. This acts as a force directed towards the interior of the particle, and thus can be considered as a pressure, stabilizing the phase with the lower molar volume, which for gallium, as stated above, is the liquid phase. Indeed, a simple calculation gives for the surface tension ca. 106 kN m^{-2}, which corresponds to an undercooling of more than 500 K, in agreement with the above findings [9]. On the other hand, nucleation theory gives, at 90 K a radius for the critical nuclei of ca. 4 nm, which is comparable with that of the nanodroplets. In other word, the nanodroplets may not be large enough to allow for nucleation, and therefore crystallization is impeded also by kinetic arguments [9].

2.4 *How Nanoparticle Size Stabilizes Different Solid Ga Phases*

As explained in Sect. 2.1, the EXAFS part of an X-ray absorption spectrum contains useful information on the radial distribution function around the photoabsorber. This can be exploited in the study of the different behaviors of Ga nanoparticles of different sizes upon cooling. In Fig. 4, the Ga K-edge EXAFS Fourier transforms are shown at different temperatures for four different nanoparticle samples with varied sizes, namely 5, 10 and 30 nm, along with that of a continuous film 150 nm thick [13].

Each of the samples shows a characteristic behavior. For the continuous film, when $T \leq 300$ K, a main peak at 2.5 Å is coupled to a second peak below 2 Å. These two peaks seem to merge into a single broad feature for $T \geq 305$ K. The continuous film constituted below 305 K is in the solid α-Ga phase. The EXAFS data analysis, not shown here for the sake of conciseness,[1] allows one to attribute the 2 Å peak to the dimeric Ga-Ga distance, while the main peak at 2.5 Å is attributed to the next nearest neighbors coordination shells: at 90 K, the data fitting gives one neighbor at 2.479(3) Å, and 6 neighbors at 2.720(4) Å, which is consistent with the structure of α-Gallium.

[1] It may be found in [13].

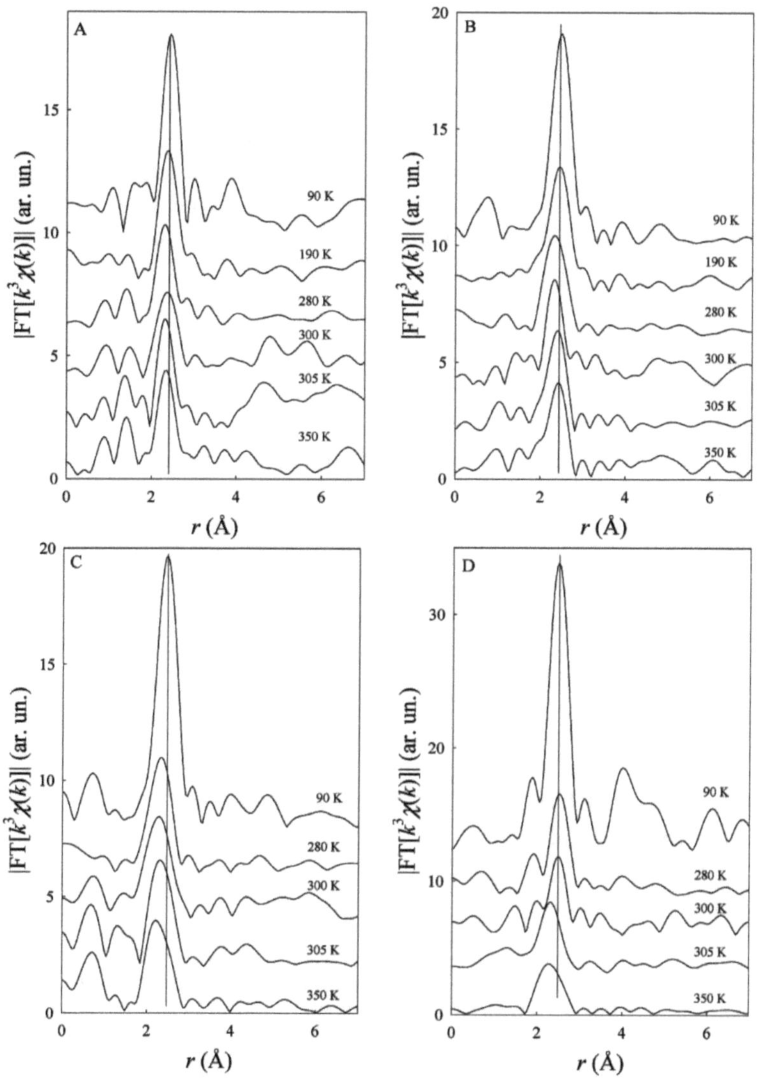

Fig. 4 Ga-K edge EXAFS Fourier at the indicated temperatures for: (**a**) 5 nm sample; (**b**) 10 nm sample; (**c**) 30 nm sample; (**d**) 150 nm sample. Reprinted with permission from [13]. (Courtesy of the original publisher. © 2007 American Chemical Society)

For the 10 and 30 nm nanoparticle samples, two different behaviors are detected. For $T \leq 190$ K the main peak in the EXAFS FT is at ca. 2.5 Å, decreasing in intensity with increasing T. Then, the main peak jumps to lower distances at $T = 280$ K, and then gradually returns to higher r with further increasing T, and shows a tail at even lower distances for $T \geq 305$ K. For the 30 nm sample, the most significant differences are that the downward jump of the main peak is more apparent, while the tail at lower r is depressed. Concerning the EXAFS data analysis, we obtain for the 30 nm sample at $T = 90$ K 0.55 neighbors at 2.594(15) Å, 2.7 neighbors at 2.745(7) Å, and 3 neighbors at 2.873(12) Å, which is a radial distribution function consistent with the structure of δ-Gallium.

Finally, for the 5 nm sample, the EXAFS FT shows a single main peak at ca. 2.5 Å, a feature that remains almost unchanged as the temperature increases, neglecting a small change in amplitude, which is again consistent with the persistence of the liquid phase at low temperatures. The EXAFS fitting at 90 K gives one neighbor at 2.48(2) Å, and 6 neighbors at 2.715(7) Å.

2.5 *Concluding Remarks*

In conclusion, these studies demonstrate how the nanostructure strongly affects the polymorphism of solid gallium and the (meta)stability range of its liquid phase. The effect of the surface free energy is, first, to favor the metallic arrangement of the δ-phase and, then, to stabilize a liquid-like phase based on dimeric molecules even at 90 K, acting in the same direction as hydrostatic pressure. It is also noteworthy to state that the Ga-Ga distance in the dimers in the liquid phase (near 2.4 Å or below) is lower than in the α solid (near 2.46 Å), pointing toward a more covalent character of the Ga-Ga bonding in the liquid than in the solid α-Ga phase.

3 Dielectric and Fröhlich Entropy Estimation of Nanoparticles

3.1 *Brief Overview*

Structural investigations carried out with XAS and electron diffraction are fruitfully integrated with dielectric measurements and the correlated Fröhlich entropy analysis. Indeed, the use of Fröhlich's theory and interpretation has been specifically developed on real experimental data for the first time with the aim of extracting from dielectric measurements a wider amount of information about the phase evolution of nanoparticles [14, 15]. In fact, the *Fröhlich entropy estimation* is able to distinctly stress the presence of melting and premelting processes in Ga nanoparticles, in a much more clear way than usual dielectric and spectroscopic measurements [10, 14]. Moreover, the FE evaluation approach is also able to provide a continuous survey of the evolution of the degree-of-order in nanoparticles as a function of temperature, that can be strictly correlated with the structural data obtained by XAS and analogous techniques [10, 13, 16–20]. In the literature, there are some works where metallic nanosystems are investigated through capacitance measurements. These studies give several details about the phase transitions, thus highlighting premelting

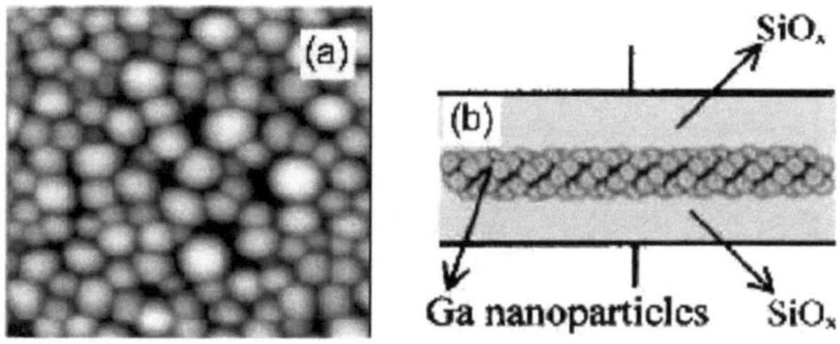

Fig. 5 Typical sample arrangement of the use of capacitance measurements and FE evaluation technique on Ga confined nanoparticles. (**a**) AFM plain-view image of the Ga nanoparticles arrays in sample; (**b**) schematic representation of the sequence SiO_x/Ga nanoparticles layer/SiO_x. Reprinted with permission from [21], Fig. 1. (Courtesy of the original publisher. ©)

and melting processes, whose dependence on the sizes of the involved particles is pointed out [10, 16, 17, 20]. The here considered experimental configurations treat samples which contain layers of Ga particles; these layers are arranged as a system of two capacitors in series, whose one layer is made of SiO_x and the second one contains the Ga clusters, as depicted in Fig. 5. In this configuration, the overall capacitance is first measured; then the contribution of the Ga layers $C = C(T)$ is extracted by a differential measurement; finally, the derivative of the Ga nanoparticle capacitance is calculated, i.e.

$$\frac{dC(T)}{dT} \propto S_E(T), \tag{4}$$

which is proportional to FE, Eqs. (9b) and (13) in chapter “Introduction: Dielectric Calorimetry and Fröhlich Entropy Estimation”. Therefore, providing the trend of dC/dT on temperature enables to analyze the behavior of Fröhlich entropy.

3.2 *Phase Transitions*

Phase transitions are studied on different samples of Ga nanoparticles. Samples with particles of 20 nm radius (Ga20) are the most representative, allowing us to highlight the features of the involved processes. Preliminary measurements as a function of temperature are carried out with transmission electron microscope (TEM), whose results are reported in Fig. 6. They pinpoint that in Ga20 sample a solid-liquid transition of the Gallium arises at $T_c = 250\,\text{K}$ [17]. The TEM technique identifies the solid crystalline phase as δ-phase, which displays a trigonal structure; this excellently agrees with the data in the literature. It is worth noting that, for the transition from solid to liquid phase, there is evidence of the melting processes arising at about 65 K below the temperature of full melting. The same samples exhibit a temperature evolution of the dielectric function which is shown

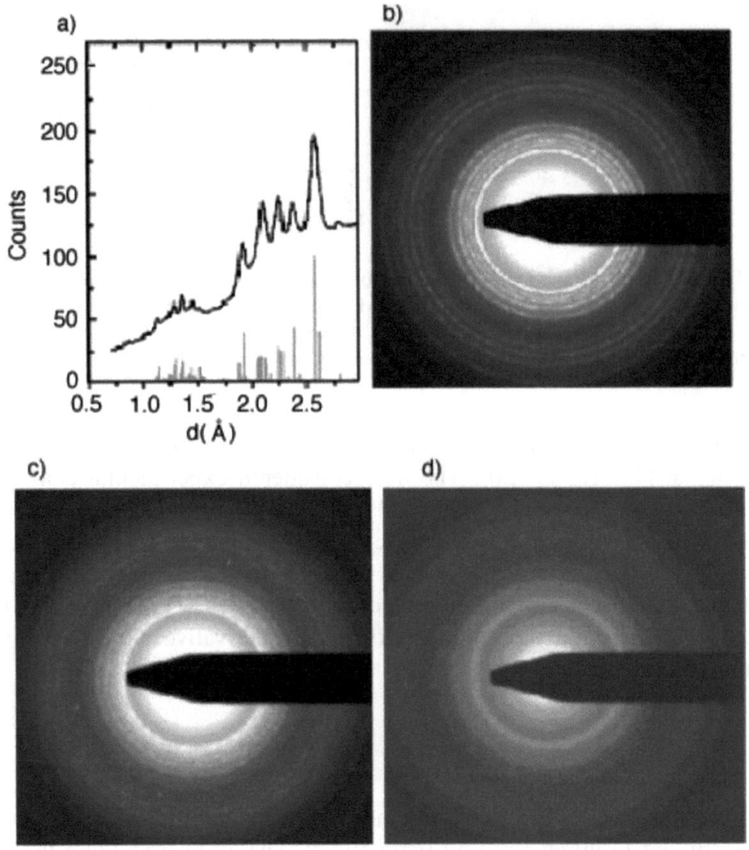

Fig. 6 TEM measurements—(**a**) Ga δ phase: selected area of electronic diffraction (SAED) radial plot of Ga clusters in the solid phase. In the upper part of the figure the calculated diffraction pattern of the Ga nanoparticles in δ phase is shown. (**b**) The correspondent SAED pattern. The SAED patterns at (**c**) 240 K and (**d**) 250 K, respectively, pointing out solid-liquid phases and full liquid phase. Reprinted with permission from [17], Fig. 3. (Courtesy of the original publisher. ©)

in Fig. 7. There, both real and imaginary parts are reported from measurements carried out around the temperature of 185 K. An evident peak is found, marking a clear transition. We underline that the TEM measurements (Fig. 6) inspect hints of a transition only at 240 K and its end at 250 K: this demonstrates that the dielectric technique and FE evaluation have a high sensitivity.

The curve of FE corresponding to the dielectric response in the Plot of Fig. 7 is shown in Fig. 8. There, FE is given as the derivative of the capacitance as a function of temperature, dC/dT. We recall that, of course, dC/dT coincides with Fröhlich entropy up to a multiplicative constant, i.e.

$$\frac{dC}{dT} \propto \frac{\partial \varepsilon}{\partial T} \propto S_E. \tag{5}$$

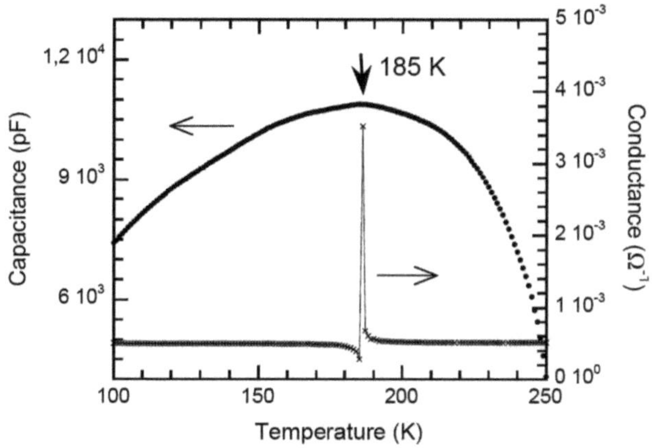

Fig. 7 20 nm size sample: capacitance and conductance of the Ga nanoparticles layer as a function of temperature in the range below the full melting temperature. Reprinted with permission from [16], Fig. 3. (Courtesy of the original publisher. ©)

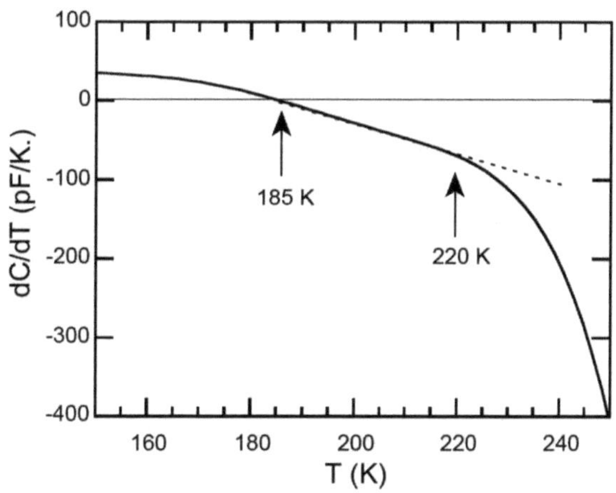

Fig. 8 Derivative of the Ga layer capacitance C of Fig. 7 as a function of the temperature T, vs. temperature (solid line). The derivative is proportional to S_E due to the applied field. The change of sign, at $T = 185$ K, points out an order-disorder transition. The dotted line shows a linear behavior in the 185–220 K interval showing the progressive disorder increasing of the atoms before inner melting starting at 220 K. Reprinted with permission from [16], Fig. 4. (Courtesy of the original publisher. ©)

By exploiting Relationships (5) and the usual interpretation of FE, we may say that the curve of dC/dT depicted in Fig. 8 pinpoints a transition from an ordered to a disordered phase, corresponding to the change from a positive to negative sign of dC/dT. This behavior is interpreted as a passage from the solid to the liquid state of Ga nanoparticles. The trend of dC/dT indicates an increasing disorder with the temperature up to 250 K. In the same Plot, we also note that at 220 K a change of slope arises, pointing out signals of pretransitional processes. These are found well before the effective occurrence of the transition at 185 K, that is indicated by the change of sign of dC/dT.

The above discussed (Sect. 2.4) comprehensive EXAFS investigations on this kind of Ga nanoparticle samples confirm in detail the polymorphism and, in particular, the presence of the pretransitional processes here stressed (Fig. 4) through dielectric measurements and FE analysis [13].

3.3 Melting Processes

In this context, the matter of melting processes needs a more detailed discussion. We recall that a specific peculiarity of nanoparticles and other low-dimensional structures is the presence of a large fraction of surface layers, whose component atoms constitute a not-negligible percentage of the total amount. In such physical systems the characteristic features exhibited by the melting processes are strongly affected by the surface atoms. Under these conditions, the use of FE evolution estimation, which is able to provide continuous information as a function of temperature, allows us to deeply analyze critical phenomena.

If we focus on the order/disorder transition, we may see that melting can be considered the most typical and extensively observed kind of phase transition, which in Ga nanoparticles has been specifically addressed in some works exploiting the FE evaluation [10, 17]. This topic is worthy of a accurate investigation because a complete satisfactory description is often difficult to be accomplished. In this vein, an challenging issue is the significant deviation of thermodynamic features shown by Ga in nanoparticle configuration with respect to the usual bulk arrangement. Actually, as we hinted above, the surface atomic layers are known to be a fundamental ingredient, whose importance was pointed out in several experimental investigations [22, 23]. Investigation techniques with high sensitivity, such as transmission electronic microscopy (TEM) and dark-field electron microscopy (DFEM), demonstrate that the phase transitions of Ga nanoparticles experience premelting processes, where liquid atomic layers form some temperature degrees below the complete melting. These liquid layers display an evolution with increasing T, with a progressive expansion, as shown in Fig. 6. The role played by surface atoms in metallic compounds was previously highlighted in two-dimensional Pb crystalline samples, where surface pre-roughening, roughening, reconstruction, and, above all, order-disorder transitions [24].

This same evolution becomes particularly relevant when such metallic systems have nanometric scales, where the amount of atoms belonging to interfaces becomes predominant. This is specifically tested on Ga nanoparticles, whose investigations points out the arising of surface changes around a solid core some temperature degrees before the accomplishment of melting [25]. A phenomenological description of such effect can be provided by analyzing the thermodynamic free energies correlated with the atomic layers. If $\gamma = \gamma(T)$ is the free energy per unit area, we may define γ_{sm}, i.e. the free energy of solid-matrix, γ_{sl}, the free energy of liquid-matrix, and γ_{lm}, free energy of solid-liquid interfaces. The melting is triggered by the balance of such quantities, so that it arises when the inequality

$$\gamma_{sm} - \gamma_{sl} - \gamma_{lm} > 0 \tag{6}$$

is satisfied. In such system, the TEM investigations highlight a complete melting of the Ga nanoparticles between 240 and 250 K, without discriminating the occurrence of premelting effects. This can be grasped in the reported Plots of selected-area electron diffraction (SAED) patterns shown in Fig. 6. On the other hand, the technique of FE evaluation applied to similar samples, whose data are reproduced in Fig. 6, clearly allow us to identify precursor phenomena of melting in the 185–220 K temperature region.

Let us analyze in detail the dielectric data. It is worth stressing that, being the dielectric response strictly correlated with the surface polarization of the nanoparticles, the measurement of $\varepsilon(T)$ and its derivative is selectively sensitive to the signal of the first atomic layers of the system. We may here note that the Plot in Fig. 8, depicting the derivative $dC(T)/dT \propto S_E(T)$, reports the FE change of sign at 185 K, thus identifying the onset of an order-disorder transition which extends to 220 K, as hinted above. The following temperature region, from 220 to 250 K, may be directly compared with TEM results of Fig. 6. In this temperature interval, the $dC(T)/dT$ curve has a progressively more marked slope, as expected, when the instabilities increase and start to involve the inner part of each nanoparticle [10, 17, 26]. Hence, such a temperature range, where the slope diverges, identifies the conditions where the melting of the inner (not surface) atomic layers occurs. Moreover, in the same Plot, the linear trend observed in the temperature range 185–220 K coincides with the range where TEM data are not modified by T-dependent processes.

FE evaluation data may be easily discussed and compared with the literature, providing a more complete and exhaustive picture. Following the data provided by [22], and confirmed by subsequent studies such as [23], the curve trend in Fig. 8 can be interpreted as the onset of vibrational instabilities at the very surface. These instabilities are known to arise before reaching a critical point, which, for melting processes, is given by the so-called *Lindemann criterion* [27–29]. This is mathematically expressed in terms of the *Lindemann ratio*, or *Lindemann parameter*, usually indicated with

$$\delta_L = \frac{\delta L_{rmsd}}{L}, \tag{7}$$

which is the ratio between the *root-mean-square displacement*, δL_{rmsd}, of a particle from its equilibrium position to the average distance between them L. Relation (7) identifies two temperature regimes: a first one where δ_L is limited, so the system is (almost) stable; a second one where δ_L diverges, pinpointing the occurrence of the melting [22, 23]. In order to apply such a criterion to our discussed data of

Ga nanoparticles, we consider the prediction of the behavior of similar physical systems. This was carried out for a paradigmatic system, a set of nanosystem made of 6912 atoms in the context of molecular dynamic (MD), obeying a Lennard-Jones potential and overheated using appropriated step-by-step calculation procedures [22, 23]. The considered size is comparable to that of our metallic Ge nanoparticles, so the comparison is fully congruent. This allows us to compare the FE evaluation data of Fig. 8 with the MD modeling data reported, from [22], shown in Figs. 9 and 10. We note that the FE (dC/dT) divergence as a function of T displays a good correspondence with the divergence of the Lindemann disordered atoms on T. The linear trend of the Fröhlich entropy variation curve in the range 185–220 K is reasonably attributed to the linear increase of the *root-mean-square displacement* of the surface atoms before reaching the region of inner melting (Figs. 9 and 10), in strict agreement with [22]. We underline that the dielectric measurements, and more so the correlated FE evaluation analysis, have a very high specific sensitivity to the evolution of the surface atoms. This represents a powerful tool for obtaining information about the different steps of pre-melting and full melting processes, which may be followed almost continuously over a wide temperature region. In this connection, the identification of two temperature regimes accomplished through the analysis of the dC/dT trend is a remarkable result, which demonstrates the potential of the analysis of Fröhlich entropy calculated from the measured dielectric response.

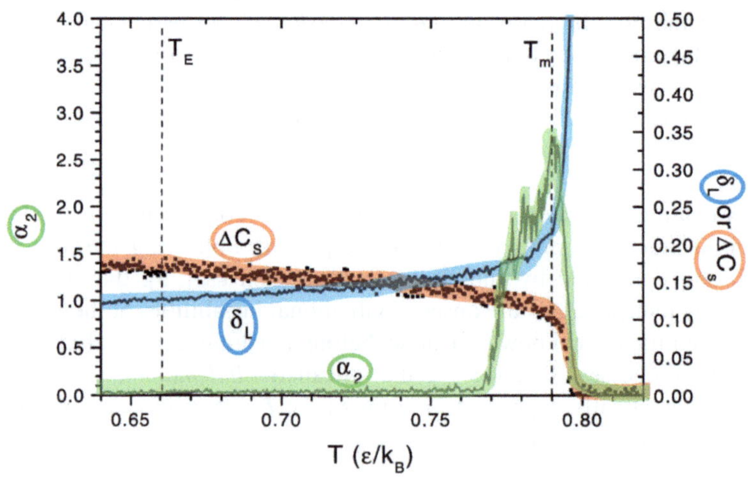

Fig. 9 The *experimental* Lindemann parameter (δ_L, blue band), the effective shear modulus ΔC_S (red band), and α_2 parameter (green band) versus temperature. Reprinted with permission and adapted from [22], Fig. 2. (Courtesy of the original publisher. ©)

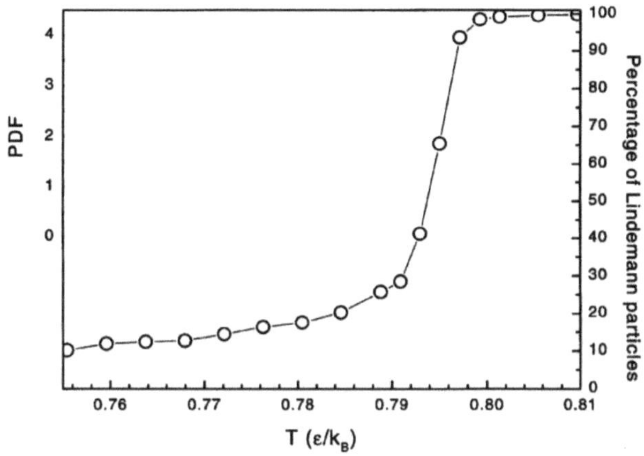

Fig. 10 *Calculated* percentage of Lindemann particles as a function of temperature. Reprinted with permission and adapted from [22], Fig. 3. (Courtesy of the original publisher. ©)

4 General Conclusions

The study of phases, phase transitions and melting processes in Ga nanoparticles is a complex topic, revealing a rich phenomenology and many interesting physical features. This can be addressed through several complementary experimental techniques. In particular, dielectric and FE evaluation techniques have a specific sensitivity to the effects of systems in nanometric and reduced-dimensionality configurations. This is due to their high sensitivity to surface effects. On the other hand, the use of synchrotron radiation enables to identify specific regimes where the overall structural melting is preceded by premelting regimes, such as e.g. the formation of dimers. However, Fröhlich Entropy evaluation experimental approach provides data in full agreement with the other standard investigation techniques, demonstrating that it is a powerful method in this kind of studies.

References

1. D. Hanaor, C. Sorrell, J. Mater. Sci. **46**, 855 (2011)
2. W. Blevin, B. Steiner, Metrologia **27**, 3 (1990)
3. F. Laves, Z. Krist Cryst. Mater. **84**, 256 (1933)
4. M. Nisoli, S. Stagira, S. De Silvestri, A. Stella, P. Tognini, P. Cheyssac, R. Kofman, Phys. Rev. Lett. **78**, 3575 (1997)
5. H. Konrad, J. Weissmüller, R. Birringer, C. Karmonik, H. Gleiter, Phys. Rev. B **58**, 2142 (1998)
6. P. Tognini, G. Parravicini, A. Stella, L. Fornari, R. Kofman, P. Cheyssac, M. Giardini, Thin Solid Films **380**, 230 (2000)
7. M. Bernasconi, G. Chiarotti, E. Tosatti, Phys. Rev. B **52**, 9988 (1995)
8. L. Bosio, H. Curien, M. Dupont et A. Rimsky, Structure cristalline de Gaδ, Acta Cryst. B **29**, 367–368 (1973)
9. G. Parravicini, A. Stella, P. Ghigna, G. Spinolo, A. Migliori, F. D'Acapito, R. Kofman, Appl. Phys. Lett. **89**, 033123 (2006)

10. G. Parravicini, A. Stella, P. Tognini, P. Merli, A. Migliori, P. Cheyssac, R. Kofman, Appl. Phys. Lett. **82**, 1461 (2003)
11. E. Søndergård, R. Kofman, P. Cheyssac, A. Stella, Surf. Sci. **364**, 467 (1996)
12. M. Feiters, R. Strange, N. Binsted, *EXCURVE* (International Union of Crystallography, Dordrecht, 2024)
13. P. Ghigna, G. Spinolo, GB. Parravicini, A. Stella, A. Migliori, R. Kofman, J. Am. Chem. Soc. **129**, 8026 (2007)
14. J. Parravicini, G. Parravicini, Res. Phys. **28**, 104571 (2021)
15. A. Jonscher, J. Phys. D Appl. Phys. **32**, R57 (1999)
16. G. Parravicini, A. Stella, M. Ungureanu, P. Merli, A. Migliori, P. Cheyssac, R. Kofman, Eur. Phys. J. D **24**, 219 (2003)
17. G. Parravicini, A. Stella, M. Ungureanu, P. Merli, A. Migliori, R. Kofman, Phys. Status Solidi B **237**, 374 (2003)
18. G. Parravicini, A. Stella, M. Ungureanu, R. Kofman, in *Proceedings - Electrochemical Society*, vol. PV 2004-13 (Electrochemical Society, Pennington, 2005), pp. 313–323. Nanoscale Devices, Materials, and Biological Systems: Fundamental and Applications - Proceedings of the International Symposium, 3 October 2004 - 8 October 2004, Conference code 66169
19. G. Parravicini, A. Stella, M. Ungureanu, R. Kofman, Appl. Phys. Lett. **85**(2), 302 (2004)
20. G. Parravicini, A. Stella, R. Kofman, Trends Vacuum Sci. Technol. **6**, 127 (2004)
21. G. Parravicini, A. Stella, M. Ungureanu, R. Kofman, Appl. Phys. Lett. **85**, 302 (2004)
22. Z. Jin, P. Gumbsch, K. Lu, E. Ma, Phys. Rev. Lett. **87**, 055703 (2001)
23. R. Cahn, Nature **413**, 582 (2001)
24. J. Frenken, P. Marée, J. van der Veen, Phys. Rev. B **34**, 7506 (1986)
25. A. Stella, P. Cheyssac, R. Kofman, P. Merli, A. Migliori, Mater. Res. Soc. Symp. Proc. **400**, 161 (1996)
26. H. Fröhlich, *Theory of Dielectrics*, 2nd edn. (Claredon Press, Oxford, UK, 1958)
27. J. Gilvarry, Phys. Rev. **102**, 308 (1956)
28. J. Hansen, Phys. Rev. A **2**, 221 (1970)
29. M. Magomedov, Solid State Commun. **391**, 115647 (2024)

Fröhlich Entropy in Soft Matter and Extreme Conditions

Paul Ben Ishai and Yuri Feldman

Abstract Fröhlich introduced his concept of dielectric configurational entropy in his monograph, The Theory of Dielectrics, in 1949. Since then, it has slowly gained traction in dielectric research. We review the basic thermodynamic concept behind it and summarize its limited application in soft condensed matter systems such as liquid crystals, simple liquids and solutions, dilute electrolytes, polymers and in biophysical settings. As examples of biological systems, we present its use to explore dipolar configuration in studies of Bacteriorhodopsin and hydrated lysozyme, respectively. We then turn to the limitations of Fröhlich's original derivation. The original derivation assumes that the permittivity has no field dependence. We rederive the entropy to account for systems where there is a distinct non-linear field response. As an example, we present a ferroelectric crystal measured at temperatures around its ferroelectric phase transition (0–40 °C) and under a bias voltage sweep (−40 to 40 V) at each temperature point. The dielectric entropy is calculated and used to trace dipolar ordering under different conditions of the imposed bias field. As we state in our conclusion, this opens new perspectives for analyzing field-induced effects and critical behaviors in functional materials and offers a thermodynamic language that is both compact and powerful.

1 Introduction

Herbert Fröhlich was a foundational figure in the theoretical physics of dielectrics and an "old time" theoretician. His classic monograph, Theory of Dielectrics [1], exemplifies this approach, with derivations that are both logically rigorous and pedagogically insightful. As a particular example, the formula derived by Fröhlich

P. Ben Ishai (✉)
Department of Physics, Ariel University, Ariel, Israel
e-mail: paulbi@ariel.ac.il

Y. Feldman
Institute of Applied Physics, Hebrew University of Jerusalem, Jerusalem, Israel
e-mail: yurif@mail.huji.ac.il

J. Parravicini (ed.), *Fröhlich Entropy Estimation and Dielectric Calorimetry*,
Advances in Dielectrics, https://doi.org/10.1007/978-3-032-21454-6_7

for the configuration entropy of dipolar system induced by an external bias field is both simple and elegant. Although attractive, it has found little application in modern dielectric studies—a tendency that may reflect the penchant for modern studies to concentrate on mechanistic interpretations of data rather than considering the energy state of the system. Nevertheless, the temperature evolution of the configurational entropy of the system offers a valuable perspective on dipolar ordering and the development of long-range dipolar correlation of the system. Complimentary methodologies such as Differential Scanning Calorimetry gloss over this point and it usually remains hidden from us. The elegance of Fröhlich's approach illuminates this hidden world.

Fröhlich's derivation was later repeated in Böttcher's Theory of Electric Polarization Volume 1 [2]. Further examinations of the thermodynamic state were added by Parravicini [3, 4]. Parravicini examined the basic assumptions made by Fröhlich in order to derive his expression for entropy. Although not clearly stated in the original monograph, Fröhlich assumed that the thermodynamic dielectric process would be isothermal and reversible, and that the material would be isotropic and linear in its response to the field. He also assumed that dielectric losses would be minimal, i.e. one can assume a quasi-static approximation for any dipolar correlation. Under these conditions his result for the configurational entropy was[1]

$$S(T, E^2) = S_0(T) + \frac{\varepsilon_0 E^2}{2} \frac{\partial \varepsilon_s}{\partial T} \tag{1}$$

Here $S_0(T)$ is the configurational entropy of the system in the absence of a field. The demand of linearity in the response to the field allows only a temperature dependence of the static permittivity, ε_s. It also implies that any dielectric measurement perturbs the system by imposing of measure of order beyond the natural order in the system as reflected by $S_0(T)$. But that, at least in terms of the state equation for the system, violates the definition of linearity. In his monograph [1] Fröhlich also left us his state equation,[2]

$$U(T, E^2) = U_0(T) + \frac{\varepsilon_0 E^2}{2} \left(\varepsilon_s + T \frac{\partial \varepsilon_s}{\partial T} \right), \tag{2}$$

Here the final terms can be recognized as the "dielectric entropy" term of Eq. (1). In complex dielectric systems, a plethora of frequency dependent processes can be recognized and even modeled by acceptable dielectric functions, such as the Debye or Cole-Cole fitting functions [5]. The statistical mechanics approach of Fröhlich to

[1] This is the same relationship as Eq. (9b) in chapter "Introduction: Dielectric Calorimetry and Fröhlich Entropy Estimation" and Eq. (33) in chapter "Elements of Thermodynamics of Dielectrics and Basic Theory of Fröhlich Entropy".

[2] This is the same relationship as Eq. (9a) in chapter "Introduction: Dielectric Calorimetry and Fröhlich Entropy Estimation" and Eq. (32) in chapter "Elements of Thermodynamics of Dielectrics and Basic Theory of Fröhlich Entropy".

link Free energy and the static permittivity can be generalized by equating ε_s to the static limit of the dynamic process. One can then consider that dielectric entity as a sub-system of the ensemble in an effective static equilibrium with its surroundings. It is then one can recognize the true gift of Fröhlich's correlation.

Normalizing the entropy term by the field strength used to probe the linear response, an idea proposed and used first by G.B. Parravicini et al. in terms of capacitance derivative [6–8] then made more explicit by Jadżyn et al. [9, 10] and later again by Parravicini [4, 11], one has an independent measure of correlation. In many ways this is stronger than considering Kirkwood's correlation factor [5, 12, 13] or Fröhlich's $B(T)$ function [1], as it is independent of the assessment of higher frequency contributions usually summed as ε_∞. In other words, the ability of the particular dipole system to correlate one with the other is proportional to $\Delta S_E(T)$, defined as

$$\Delta S_E(T) = \frac{S(T, E^2) - S_0(T)}{E^2} = \frac{\varepsilon_0}{2}\frac{\partial \varepsilon_s}{\partial T} \tag{3}$$

This quantity is independent of the field strength and reflects only material characteristics. Jadżyn et al. applied Eq. (3) to explain the evolution of the Isotropic to Nematic (I-N) phase transition of $C_nH_{2n+1}C_yH_xPhN{=}C{=}S$ (for n = 6, 4- (trans-4'-n-hexylcyclohexyl) isothiocyanatobenzene (6CHBT)) liquid crystals [10]. They demonstrated that on cooling from 75 °C to the transition temperature at 43 °C ΔS_E increases from a negative value of −0.05 (in units of K^{-1}) to almost zero at the transition temperature. This result demonstrates nicely the use of ΔS_E as an effective order parameter. They then applied the same reasoning to Binary mixtures of 6CHBT and an admixture non-polar compound $C_2H_5C_yH_xPhC_9H_{19}$,$2C_yPh_9$ [9] In this case they noted a shifting of the I-N transition temperature as a function of the admixture concentration. ΔS_E also reveals the modification to the order parameter as the polar 6CHBT is steadily diluted.

Another set of liquid systems where ΔS_E could offer insight is the evolution of dipolar ordering Liquid Lactones, such as liquid γ-butyrolactone (GBL) and γ -valerolactone (GVL). Świergiel and Jadżyn [14] exploited Eq. (3) to illustrate in a simple liquid how the phenomenological behavior of the permittivity as reciprocal to the temperature, $\varepsilon_s \sim 1/T$ could be adequately explained by a loss of correlation as the temperature increases. A more comprehensive definition of the temperature dependence of the permittivity, $\varepsilon_s \sim (gN\mu^2)/(K_BT)$, [1] hints that in this case, thermal decorrelation dominates over induced dipolar ordering. Here g is the correlation factor, N the dipole density, μ the permanent dipole moment, K_B Boltzmann's constant and T the temperature, in this case the variation behaves as $\Delta S_E(T) \simeq -1/T^2$. Świergiel et al. also considered ordering of dilute electrolytes, such as Lithium Perchlorate ($LiClO_4$) in Polyethylene Glycols [15] and revealed similar behavior of a steady degrading of dipolar correlation as the samples were heated. Ferroelectric polymer systems were studied by Kochervinskii et al. [16] and here too Eq. (3) was used to clarify the nature of long range dipolar ordering.

Although Eq. (1) was published in 1949 [1] further theoretical development had to wait until 2016 when Matyshov considered its application in glass formers [17]. Following a suggestion by Johari [18] to use strong electric fields to probe the glass transition, Matyshov astutely noted that Fröhlich's entropy is thermodynamic and should not as such contribute to the configurational entropy of the system. His argument was that configuration entropy enumerates the number of available energy states in the system and the effect of the field should be to redistribute the population among these states rather than to add to them. He also noted that there should be a non-linear component involved and that this is not accounted for. However, he did not take these considerations further. One can argue against his position by observing that the degree of freedom lost by imposition of dipolar order by the electrical field denies the system access to available energy states and that this most definitely effects the configurational entropy.

Further theoretical work was done by Parravicini [4] in considering anisotropic systems. While less relevant for traditional soft condensed matter systems, this approach comes to its own when applied to ferroelectric crystals. His derivation implies coupling between different directions of the crystal principle axes and Eq. (1) must be modified to read[3]

$$S(E,T) = S_0(T) + \frac{1}{2}\sum_{ij}\left(E_i \frac{\partial \varepsilon_{ij}}{\partial T}\right) E_j. \tag{4}$$

For soft condensed matter systems this idea is important if anisotropy is introduced. A case in point is again ferroelectric polymers. Kochervinskii and Malyshkina [19] revisited VDF/TrFE 70/30 thin films. However, with the strong coercive fields they imposed on the films (~ 0.1 MV/m), they effectively suppressed any off-diagonal elements of the dielectric tensor and so Eq. (1) was sufficient for analysis.

In the following sections we will consider 3 systems: the first two illustrate the insights one can gain by applying ΔS_E to biological systems and the third example extends Fröhlich's original idea to non-linear regimes of the dielectric permittivity. In order to explore non-linear effects, we return to inert ferroelectric crystals as a model system. We start with membranes of Bacteriorhodopsin, a bacterial membrane protein [20, 21].

2 Bacteriorhodopsin Membranes and Ferroelectric-like Behavior

Bacteriorhodopsin (bR) is a membrane protein found in the bacterium Halobacterium Salinarium. The name is a misnomer as the animal is not a true bacteria, but a member of the Archæa domain [22]. These are single celled micro-organisms capable of living in extreme high salinity environments. Retinal, a vitamin A

[3] This is the same relationship as Eq. (9) in chapter "Generalized Fröhlich Entropy: Anisotropic and Nonlinear Media".

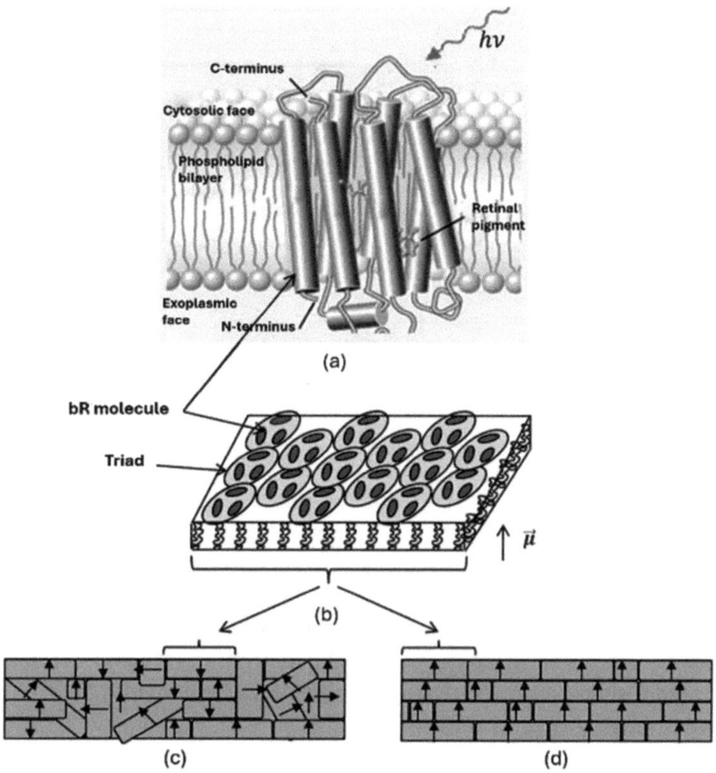

Fig. 1 From (**a**) a schematic representation of a single bR protein; (**b**) the triad of bR molecules in the membrane; (**c**) non-orientated and (**d**) orientated bR films. Reproduced with permission from [5]. (Courtesy of the original publisher Oxford University Press ©)

derivative, is covalently bonded to bR and is responsible for the purple coloring of these membranes. The protein bR acts as a light driven proton pump, changing the proton concentration profile across the membrane, something that is converted to chemical energy [23]. Approximately 70% of the membrane is bR with the rest provided by lipids. The protein consists of 7 closely packed α-helices that extend perpendicular to the membrane face. Three bR macromolecules are assembled into dense triads with lipids separating the molecules. This assembly leads to highly regular lattices [24]. The structure is illustrated in Fig. 1. Integral to its action is the motion of dipolar groups in the molecule. Consequently, this was investigated using Time Domain Dielectric Spectroscopy [25] in the frequency range 500 kHz to 1 GHz [21]. Orientated and Non-orientated membranes were prepared from centrifuged cultivated Halobacterium. Further centrifugation separates the purple membrane fraction, which can then be deposited on a substrate or electrode. The full details of preparation can be found in Refs. [26, 27]. Two groups of samples were prepared, one unoriented and the other oriented by the application of a 10 V

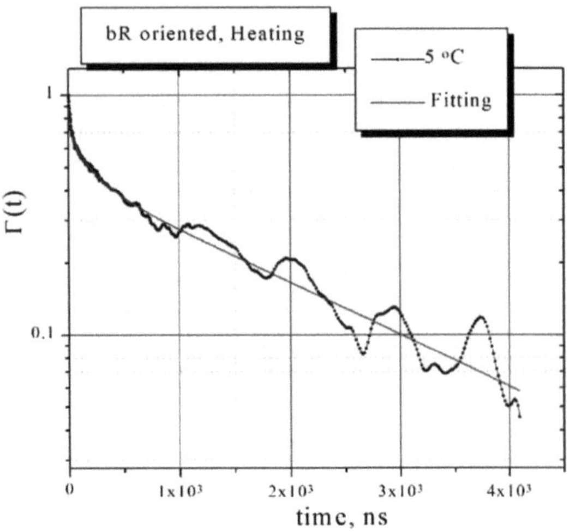

Fig. 2 A typical dipole correlation function derived from a TDDS measurement of bR membranes. The measurement is of the orientated membrane at 5 °C. The black symbols are the experimental data and the red line is the fitting function

dc voltage during sedimentation (see Fig. 1c and d). Each film was approximately 30 μm thick. The fragments have a permanent dipole moment of $\mu \sim 2.5 \cdot 10^6$ D, while the bR molecule and retinal have moments of $\mu_{bR} \sim 60 - 98$ D and $\mu_R \sim 10$ D respectively [20]. The macroscopic dipole correlation function, $\Gamma(t)$, is derived from the signal shape recorded during the TDDS measurement [25] and is related to the permittivity by Fourier transform,

$$\Gamma(t) = \frac{\langle \vec{M}(0)\vec{M}(t)\rangle}{\langle \vec{M}(0)\vec{M}(0)\rangle}, \qquad \varepsilon^*(\omega) = \varepsilon'(\omega) - i\varepsilon''(\omega) = \mathcal{L}[\Gamma(t)], \tag{5}$$

where $\mathcal{L}[\,]$ is the Fourier transform operator, $\vec{M}(t)$ is the macroscopic dipole moment of the sample at time t and the brackets signify thermodynamic averaging. A typical result is shown in Fig. 2. The data was modeled using the sum of 3 exponential decay functions with differing time constants and amplitudes. The first and slowest process has a relaxation time of around 1 μs and is attributed to anisotropic rotation of the bR molecule as a whole.

The second process reflects anisotropic vibration of the α-helices in the membrane and the final process was ascribed to a mixture of anisotropic motion of the retinal, the mobility of protein polar groups and lipid heads [20, 21]. We concentrate on the first process in this monograph. Differential scanning calorimetry of the orientated membrane reveals a small endothermic process occurring at 18 °C, attributed to the freezing of lipid chains in the membrane, which is not noted in the DSC of the non-orientated sample [21]. Additionally, the dielectric strength of the first process exhibits a hysteresis between successive heating and cooling runs (Fig. 3). The crossover between them also occurs at 18 °C. Finally, when the orientated sample is measured under a bias field the dielectric strength, $\Delta\varepsilon$, displays a strong dependence on both the bias field strength and the field direction.

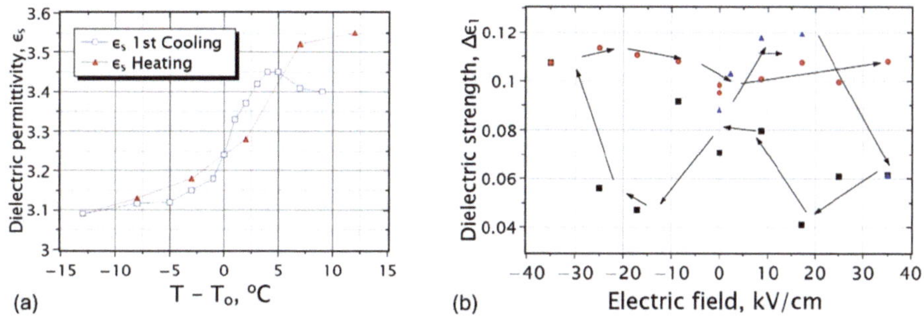

Fig. 3 (**a**) The dielectric strength of process 1, attributed to the libration of bR proteins in the membrane fragments, for the orientated sample. The temperature scale is referenced to $T_0 = 18\,°\mathrm{C}$, and (**b**) the strong hysteresis noted in the measurement of the sample under a bias voltage. The measurement protocol was 0 to +35 V (blue triangles), +35 V to −35 V (black squares) and −35 V to +35 V (red circles)

Fig. 4 The field normalized Fröhlich entropy of the orientated bR membrane. The sample was cooled from room temperature (25 °C) to 10 °C and then reheated to room temperature. ΔS_E is an effective measure of the dipole correlation and can be considered as an order parameter

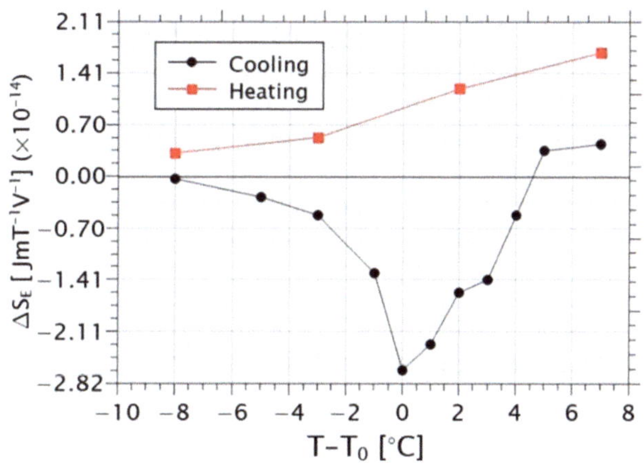

Together this suggests that bR proteins embedded in these orientated membranes act cooperatively, promoting the freezing of lipids in the membrane and hence the DSC signal, in a manner reminiscent of a ferroelectric-like phase transition.

It is instructive to calculate Fröhlich's entropy for this system. Using the definition for ΔS_E as defined in Eq. (3) we can estimate the degree of dipole correlation. The result is shown in Fig. 4. As the field strength is effectively fixed the increase or decrease in the entropy reflects the dipole correlation and can be considered as an order parameter. One notes the negative values indicate that the entropy is being reduced and so dipole order must increase. The order in the system reaches a maximum at $T_0 = 18\,°\mathrm{C}$ already recognized as the effective freezing temperature for lipids in the membrane. This implies that the interaction between the ordered dipoles of bR proteins in the membrane leading to greater correlation is "frozen in".

The interpretation of ΔS_E as an effective order parameter is further strengthened by a naïve dimensional analysis. Using the dimensions for entropy, $[\Delta S] = ([J])/([K])$ where $[J]$ signifies Joules and $[K]$ Kelvin, and for electric field strength, $[E^2] = [N^2]/[C^2]$ where $[C]$ is for Coulomb and $[N]$ is for Newton, on has

$$[\Delta S_E] = \frac{[J]}{[K]} \cdot \frac{[C^2]}{[N^2]}. \tag{6}$$

This can be further reduced by using the definition for work as the force multiplied by distance so that $[J] = [N][m]$, where $[m]$ signifies meters. The final dimensionality is then

$$[\Delta S_E] = \frac{[C^2][m^2]}{[K][J]}. \tag{7}$$

As $[C^2][m^2]$ can be recognized as the natural units for the absolute value of the dipole moment, it is clear that the ΔS_E is the rate of change in the macroscopic dipole moment of the sub system represented by the dielectric process as a function of temperature and orientating electric field.

On heating the sample ones notes that the imposed order is steadily decreased, indicated by positive values of the Fröhlich entropy. This is akin to the persistence of a coercive field in traditional hysteresis $P - E$ curve for ferroelectric crystals.

3 Application of Fröhlich Entropy to Hydrated Water in Lysozyme Powders

Building upon the use of Fröhlich's configurational entropy as a diagnostic for dipolar correlation in complex biological systems, we now turn to another biologically significant dipolar system: hydrated lysozyme powders. These systems, offer a rich testbed for exploring the dipole dynamics of confined water under biomolecular constraints.

Hydrated lysozyme powders were created from chicken egg-white by isopiestic equilibration [28] using saturated solutions of CsCl (sample 1) and NaCl (sample 2). The solutions and the lysozyme powder were kept in a desiccator at room temperature for 3 days. The calculated water mass fractions, $h = M_w/M_p$, where M_w and M_p are the mass of the water and protein respectively, are $h = 0.16$ (sample 1) and $h = 0.28$ (sample 2). Broadband dielectric measurements reveal a complex relaxation environment [29]. The samples were quenched to 130 K and measurements made on heating to 200 K. We concentrate on one process only that has been identified as the dielectric relaxation of interfacial water on the protein surface and inside the protein's active pore. Differential Scanning Calorimetry measurements reveal distinct exothermic peaks that can be attributed to the freezing

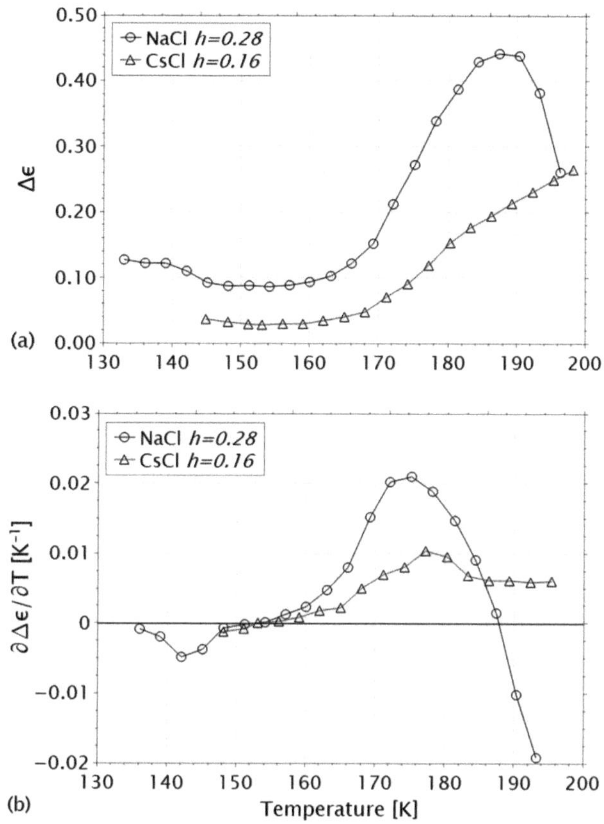

Fig. 5 (**a**) The dielectric strength for the processes associated with interfacial water in lysozymes hydrated powders. (**b**) the derivative of the dielectric strength with temperature

of the same interfacial water. Two samples were measured on cooling from 220 to 130 K, each with a hydration level of $h = 2.0$. One sample was allowed to anneal for 10 minutes at 200 K. A distinct exothermic process is evident as the annealed sample passes through 185 K. A further process is noted in both samples around 152–160 K. These critical temperature points coincide with the changes in ΔS_E, as seen in Fig. 5b. As the dielectric measurements were made on heating and $\Delta\varepsilon$ is reversible, the appropriate temperature change direction is negative in order to compare to the results of DSC. In this case ΔS_E, indicates higher dipolar ordering as the temperature is lowered towards 185 K, in keeping with the pore water crystallization noted in Fig. 6. The DSC scans also reveal that annealing is necessary for crystallization to happen, and one must assume that without it, water in the pore remains super cooled. Intriguingly, ΔS_E reveals that a partial process for dipolar ordering is occurring in the low humidity sample too (the red trace in Fig. 5b).

In this context, Fröhlich's entropy model provides a thermodynamic lens, through which it is possible to interpret these relaxation phenomena. Specifically, by tracking the temperature dependence of the static dielectric permittivity ε_s one can infer

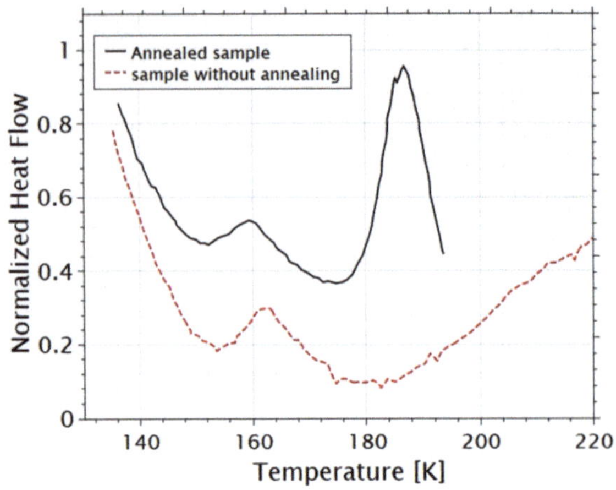

Fig. 6 Normalized Heat Scans of hydrated lysozyme powders. Two exothermic processes are visible in the annealed sample. These are associated with the freezing of interfacial water in the active pore (∼185 K) and on the lysozyme surface (∼160 K)

the configurational entropy changes associated with dipole ordering and disordering in the hydration shell. As seen in the bR system, a negative derivative suggests increasing dipole correlation (entropy reduction), while a positive slope indicates disordering. In hydrated lysozyme powders, this derivative reveals an initial order-to-disorder transition at ∼155 K, followed by a disorder-to-order transition around ∼186 K on cooling down. Other authors have noted the transition at 155 K but associated it with a glass transition rather than a crystallization [30, 31]. However, Khodadadi did note a structural relaxation of the protein that froze at $T = 180$ K [31], lending credence to the conclusions of the Fröhlich approach. This phase transition, appearing upon cooling of the samples, can be related to the reversible transition of water from the disordered to the ordered state, as noted by others [32].

Thus, Fröhlich's configurational entropy, originally derived under idealized assumptions, finds renewed relevance in the study of confined biological hydration, capturing the thermodynamic essence of dipolar interactions in hydrated proteins such as lysozyme, much like it did in the ferroelectric bR system. This further strengthens the case for adopting entropy-based frameworks in dielectric analyses of soft biological matter under extreme conditions.

4 The Extension of Fröhlich's Entropy

It is tempting to consider what the entropy picture would be if one allows the possibility that the permittivity under certain conditions would depend on the electric field as well. This is realistic scenario for ferroelectrics close to their ferroelectric phase transition. It is possible to rework Fröhlich's original idea to incorporate this dependence. In this case, the basic thermodynamic energy state equation, $U = TS + (E \cdot D)/4\pi$, leads to the following differential equation

$$dU = TdS + \frac{\varepsilon_0}{2}\left(\varepsilon_s + 2\sqrt{E^2}\frac{\partial \varepsilon_s}{\partial E^2}\right)dE^2 + \varepsilon_0 E^2 \frac{\partial \varepsilon_s}{\partial T}dT. \tag{8}$$

Inverting Eq. (8) on has

$$dS = \frac{1}{T}\left[dU - \frac{\varepsilon_0}{2}\left(\varepsilon_s + 2\sqrt{E^2}\frac{\partial \varepsilon_s}{\partial E^2}\right)dE^2 - \varepsilon_0 E^2 \frac{\partial \varepsilon_s}{\partial T}dT\right]. \tag{9}$$

One can then follow Fröhlich's original derivation, noting that the entropy of the system must be a function of 2 parameters and a total differential, so its partial derivatives must obey $\partial/\partial T(\partial S/\partial E^2) = \partial/\partial E^2(\partial S/\partial T)$. Explicitly the partial derivatives of the Entropy are,

$$\frac{\partial S}{\partial T} = \frac{1}{T}\left(\frac{\partial U}{\partial T} - \varepsilon_0 E^2 \frac{\partial \varepsilon_s}{\partial T}\right), \tag{10}$$

$$\frac{\partial S}{\partial E^2} = \frac{1}{T}\left[\frac{\partial U}{\partial E^2} - \frac{\varepsilon_0}{2}\left(\varepsilon_s + 2\sqrt{E^2}\frac{\partial \varepsilon_s}{\partial E^2}\right)\right]. \tag{11}$$

Carrying out the required derivatives and equating them results in the following relationship

$$\frac{\partial U}{\partial E^2} = \frac{\varepsilon_0}{2}\left(\varepsilon_s + 2\sqrt{E^2}\frac{\partial \varepsilon_s}{\partial E^2} + T\frac{\partial \varepsilon_s}{\partial T}\right) - T\varepsilon_0 E^2 \frac{\partial^2 \varepsilon_s}{\partial T \partial E^2}. \tag{12}$$

The final term of Eq. (12) represents pyroelectric effects in the system. Feeding Eq. (12) back into Eq. (11) results in

$$\frac{\partial S}{\partial E^2} = \frac{\varepsilon_0}{2}\frac{\partial \varepsilon_s}{\partial T} - \varepsilon_0 E^2 \frac{\partial^2 \varepsilon_s}{\partial T \partial E^2} \tag{13}$$

or

$$\frac{\partial S}{\partial E^2} = \frac{\varepsilon_0}{2}\frac{\partial}{\partial T}\left(\varepsilon_s - 2E^2\frac{\partial \varepsilon_s}{\partial E^2}\right). \tag{14}$$

As the temperature and the electric field are independent variables, the order of integration and differentiation can be exchanged. Integrating by E^2 then gives

$$S(E^2, T) = S_0(T) + \frac{\varepsilon_0}{2}\frac{\partial}{\partial T}\int\left(\varepsilon_s - 2E^2\frac{\partial \varepsilon_s}{\partial E^2}\right)dE^2, \tag{15}$$

or in terms of the electric field

$$S(E, T) = S_0(T) + \varepsilon_0\frac{\partial}{\partial T}\int\left(\varepsilon_s - E\frac{\partial \varepsilon_s}{\partial E^2}\right)EdE, \tag{16}$$

As in the original derivation, $S_0(T)$ is the configurational entropy in the absence of an electric field. If there is no dependence of the permittivity on the electric field, then the original entropy equation is regained.

Equation (16) is the basic tool that we can exploit to probe Entropy close to the ferroelectric phase transition. To illustrate this point we consider a ferroelectric crystal, engineered such that its phase transition is close to room temperature. We measured a Potassium Niobate Tantalate ($KTa_{1-x}Nb_xO_3$) perovskite crystal, lightly doped with Copper in a narrow temperature regime (0–40 °C) spanning the ferroelectric phase transition at 22 °C and under bias voltages at each temperature point from −40 to 40 V. Given the crystal dimensions this is equivalent to bias fields of −36.4 to 36.4 kV/m. The crystal was grown using the top-seeded solution growth method [33, 34]. The Ta/Nb ratio was estimated by electron microprobe analysis and was found to be approximately 62/38 per mole. The Cu doping level was 2% in the flux yielding approximately 1.5×10^{-4} per mole in the grown crystal. Samples of $1{\times}1{\times}2\ \mathrm{mm}^3$ were cut from the grown bole along the crystallographic [001] axes. The $x - y$ faces of the samples (perpendicular to the growth direction) were coated with gold electrodes to ensure an Ohmic contact. The measurements were made using a Novocontrol BDS 80 dielectric spectrometer, based on an Alpha Impedance analyzer (Novocontrol Technologies GmbH, Germany). The temperature scan was from 0 to 40 °C. At each temperature point the bias voltage was set from −40 to 40 V with a frequency sweep from 1 Hz to 10 MHz made at each point. The permittivity for all temperatures and bias voltages is shown at 1 Hz and at 10 MHz in Fig. 7. The temperature range chosen spans the ferroelectric phase transition temperature at $T_0 = 24.8\,°\mathrm{C}$, for the cubic paraelectric phase to the tetragonal ferroelectric phase [34–36]. Figure 8a shows the temperature dependence at 0 V bias of the permittivity at 1 Hz and at 10 MHz, while Fig. 8b shows the bias dependence of the permittivity at 1 Hz for $T = 0\,°\mathrm{C}$ and for $T = 24\,°\mathrm{C}$. It is clear that there is a significant effect of the bias field on the permittivity. It is therefore necessary to use Eq. (16) in order to estimate the entropic element in the phase transition. In order to implement Eq. (16)

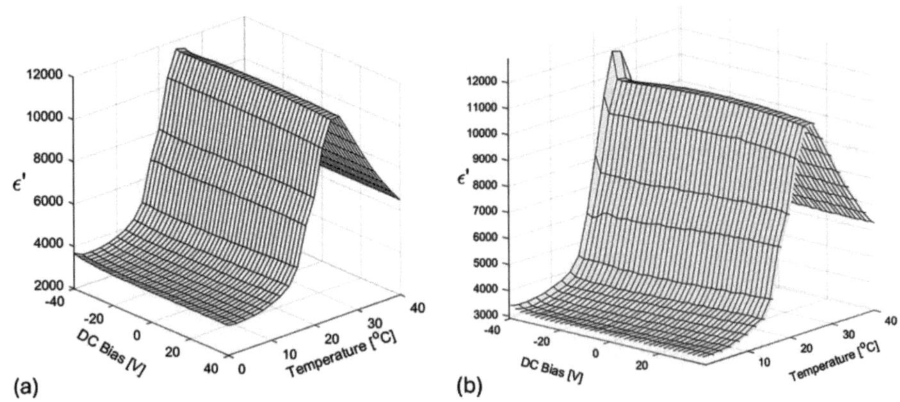

Fig. 7 The dielectric permittivity for a KTN crystal as a function of DC Bias and Temperature. (**a**) at the frequency 1 Hz and (**b**) at the frequency of 10 MHz

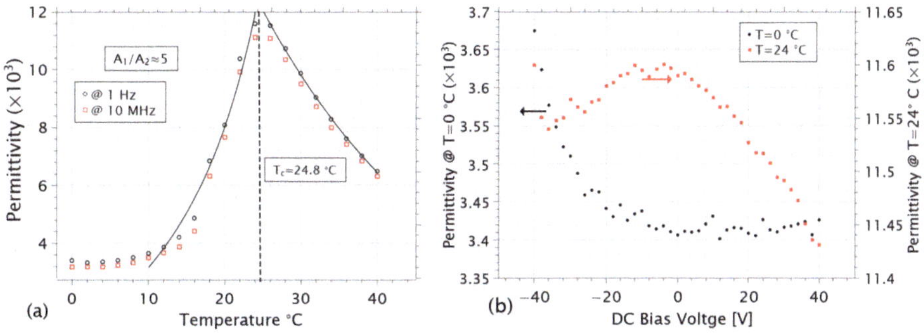

Fig. 8 (**a**) the temperature dependence of the permittivity with no bias Voltage, get for $f = 1$ Hz and 10 MHz. The lines are Curie-Weiss fits and the ratio of fitting amplitudes $A_1/A_2 = 5$ consistent with a second order structural phase transition [37]. (**b**) The permittivity as a function of bias voltage for $T = 0\,°C$ and $T = 2\,°C$, measured at 1 Hz. The field dependence of the permittivity is clearly visible

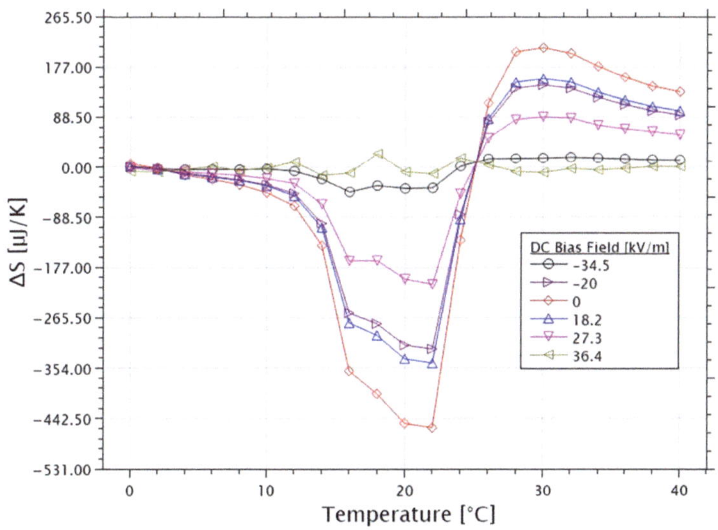

Fig. 9 The Fröhlich Entropy for a KTN crystal under bias voltages, as calculated by Eq. (16). All curves cross at the ferroelectric transition temperature $T_0 = 24.8\,°C$

a Matlab application was made. For a specific frequency the integrand of Eq. (16) was calculated from V = −40 V to the desired bias voltage up to V = 40 V, for each temperature point. Then the temperature derivative of these results was taken. The final curves are plotted in Fig. 9 for 6 bias voltage endpoints. It is immediately clear that the original supposition by Fröhlich of reversibility in the permittivity is no longer applicable. The initial bias field, in this case −36.4 kV/m, dictates the initial dipolar ordering and hence the development of ΔS_E, as defined by Eq. (3). In Fig. 9 one notes that the effect of a strong starting field is to suppress the temperature dependence of the entropy. The starting temperature of the experiment is such that the crystal is in a tetragonal ferroelectric state, with the Niobium ion in the center of the perovskite structure offset along the c-axis [36]. As the temperature is raised towards the ferroelectric phase transition ΔS_E becomes more negative as dipolar order is increased. This tendency is most acute for the zero bias field endpoint, as $S_0(T)$ dominates the transition behaviour. The Fröhlich entropy is zeroed for all bias endpoints at the ferroelectric phase transition temperature, $T_0 = 24.8\,°\text{C}$. Once into the paraelectric phase one notes that ΔS_E becomes positive, indicating a measure of disorder. Again, the tendency is strongest for the zero bias field endpoint. It is instructive to plot the maximum and minimum values of ΔS_E as a function of the bias field endpoint. This is presented in Fig. 10. These curves are clearly quadratic in nature and the parameters of the fitting are presented in Table 1. As expected, the behavior maintains the E^2 motif, but the linear term indicates a departure from Eq. (1) and the adoption of Eq. (16) as the description of the evolution of Fröhlich's configurational entropy.

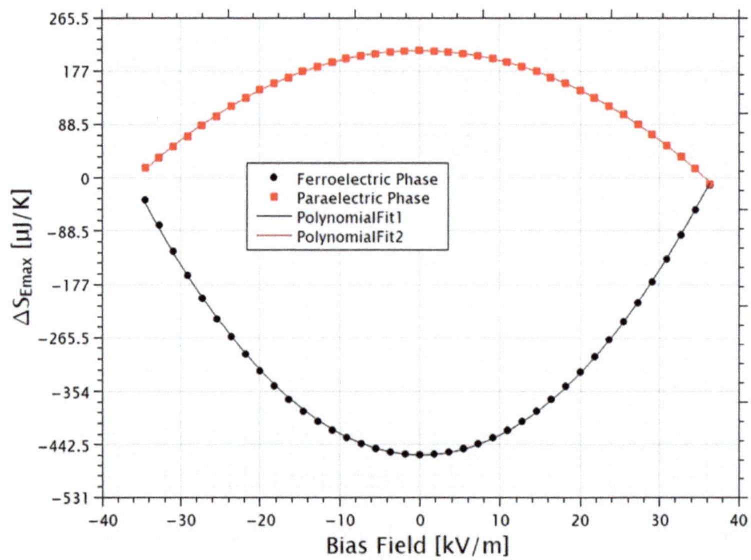

Fig. 10 The maximum (paraelectric phase) and minimum (ferroelectric phase) of ΔS_E as a function of the bias field. The fitting lines are quadratic in both cases

Table 1 The fitting parameters for the quadratic fits ($y = A_0 + A_1 x + A_2 x^2$) of the curves in Fig. 10

Phase	A_0	A_1	A_2
Ferroelectric	$(-5.190 \pm 0.006) \cdot 10^7$	$(-1.7 \pm 0.2) \cdot 10^4$	$(3.91 \pm 0.01) \cdot 10^4$
Paraelectric	$(2.382 \pm 0.002) \cdot 10^7$	$(8.07 \pm 7.04) \cdot 10^2$	$(-1.865 \pm 0.004) \cdot 10^4$

5 Conclusions

The concept of configurational entropy introduced by Herbert Fröhlich, though originally developed in the context of idealized reversible dielectric systems, reveals a remarkable and unexpected versatility when applied to real-world complex materials. In this work, we have demonstrated that Fröhlich's elegant formalism is not only of historical value but continues to offer deep insights into the fundamental behavior of dipolar systems across disciplines.

Our investigation of bacteriorhodopsin membranes and hydrated lysozyme powders has shown that the entropy-based approach provides a rigorous, physically grounded measure of dipolar ordering-capable of capturing cooperative phenomena, phase-like transitions, and dynamic correlations that elude more conventional methods. In biological systems especially, where disorder and confinement often mask thermodynamic signatures, Fröhlich's entropy serves as an effective order parameter-revealing the underlying logic of interaction and self-organization.

The theoretical generalization of Fröhlich's model to include nonlinear dielectric responses, and its successful application to ferroelectric perovskites near their phase transition, demonstrates that this approach is not limited to the linear regime. It opens new perspectives for analyzing field-induced effects and critical behavior in functional materials, and offers a thermodynamic language that is both compact and powerful.

For biophysics, this work suggests that dipolar correlation and entropy deserve more central attention and not as mere secondary characteristics. For condensed matter physics, it offers a refined toolset for interpreting order-disorder phenomena, where fluctuation, memory, and field sensitivity are deeply entangled. In short, the entropy formulation proposed by Fröhlich, long considered a theoretical curiosity, reveals itself as a profound and unifying concept. It invites us to view dielectric systems not only through the lens of response functions, but also as thermodynamic ensembles striving for coherence under constraints. This, perhaps, is the lasting legacy of Fröhlich's thinking: to uncover simplicity in complexity through the quiet rigor of theoretical physics.

Acknowledgments The authors would like to thank Prof. Aharon Agranat and Dr. Gilad Orr for KTN crystal used in these experiments.

References

1. H. Fröhlich, *Theory of dielectrics: Dielectric Constant and Dielectric Loss*, 1st edn. Monographs on the Physics and Chemistry of Materials (Oxford University Press, Oxford, UK, 1949)
2. C. Böttcher, *Theory of Electric Polarization*, 2nd edn. (Elsevier Science, Amsterdam, 1973)
3. J. Parravicini, G. Parravicini, Res. Phys. **28**, 104571 (2021)
4. J. Parravicini, Phys. B **541**, 54 (2018)
5. V. Raicu, Y. Feldman, *Dielectric Relaxation in Biological Systems: Physical Principles, Methods, and Applications* (Oxford University Press, Oxford, UK, 2015)

6. G.B. Parravicini, A. Stella, M. Ungureanu, P. Merli, A. Migliori, R. Kofman, Phys. Status Solidi B **237**, 374 (2003)
7. G.B. Parravicini, A. Stella, P. Tognini, P. Merli, A. Migliori, P. Cheyssac, R. Kofman, Appl. Phys. Lett. **82**, 1461 (2003)
8. G.B. Parravicini, A. Stella, M. Ungureanu, P. Merli, A. Migliori, P. Cheyssac, R. Kofman, Eur. Phys. J. D **24**, 219 (2003)
9. J. Jadżyn, U. Sokołowska, G. Czechowski, J. Phys. Chem. B **112**(23), 7022 (2008). https://doi.org/10.1021/jp801504f
10. J. Jadżyn, G. Czechowski, J.L. Déjardin, J. Phys. Chem. B **112**(16), 4948 (2008). https://doi.org/10.1021/jp7112494
11. A. Sassella, D. Braga, M. Campione, T. Ciabattoni, M. Moret, J. Parravicini, GB. Parravicini, J. Appl. Phys. **109**, 013529 (2011)
12. L. Onsager, J. Am. Chem. Soc. **58**, 8 (1936)
13. F. Kremer, A. Schönhals (eds.), *Broadband Dielectric Spectroscopy* (Springer, Berlin, Heidelberg, 2003)
14. J. Świergel, J. Jadżyn, Int. J. Thermophys. **33**, 783 (2012). https://doi.org/10.1007/s10765-012-1189-x
15. J. Świergiel, I. Płowaś, J. Grembowski, J. Jadżyn, J. Chem. Eng. Data **60**, 3588 (2015)
16. V. Kochervinskii, S. Bedin, I. Razumovskaya, M. Shcherbina, M. Buzin, I. Malyshkina, A. Prazdnichnyi, A. Pavlov, O. Vasilevskii, Polym. Sci. Ser. A **58**, 345 (2016)
17. D.V. Matyushov, J. Chem. Phys. **145**, 034504 (2016)
18. G.P. Johari, J. Chem. Phys. **138**, 154503 (2013)
19. V. Kochervinskii, A. Malyshkina, *9 - Dielectric Spectroscopy of Ferroelectric Polymers.* Woodhead Publishing Series in Electronic and Optical Materials (Woodhead Publishing, 2022), pp. 263–355. https://doi.org/10.1016/B978-0-12-821551-7.00001-4
20. I. Ermolina, Y. Hayashi, V. Raicu, Y. Feldman (eds.), *Proteins in Solutions and Natural Membranes* (Oxford University Press, Oxford, UK, 2015) [Online]. Available: https://www.oxfordscholarship.com/view/10.1093/acprof:oso/9780199686513.001.0001/acprof-9780199686513-chapter-11. Accessed 02 Oct 2019
21. I. Ermolina, A. Strinkovski, A. Lewis, Y. Feldman, J. Phys. Chem. B **105**, 2673 (2001). https://doi.org/10.1021/jp001054y
22. S. DasSarma, Am. Sci. **95**, 224 (2017). https://doi.org/10.1511/2007.65.224
23. D. Voet, J.G. Voet, *Biochemistry*, 4th edn. (Wiley, 2010)
24. Y. Shen, C. Safinya, K. Liang, A. Ruppert, K. Rothschild, Nature **366**, 48 (1993). https://doi.org/10.1038/366048a0
25. Y.F. et al., Rev. Sci. Instrum. **67**, 3208 (1996)
26. D. Oesterhelt, W. Stoeckenius, *Isolation of the Cell Membrane of Halobacterium halobium and its Fractionation into Red and Purple Membrane* (Academic Press, New York, 1974), pp. 667–678. https://doi.org/10.1016/0076-6879(74)31072-5
27. G. Váró, K. Bryl, Biochim. Biophys. Acta (BBA) Bioenerg. **934**, 247 (1988). https://doi.org/10.1016/0005-2728(88)90188-0
28. P. Poole, J. Finney, *Solid-phase Protein Hydration Studies* (Academic Press, New York, 1986), pp. 284–293. https://doi.org/10.1016/0076-6879(86)27023-8
29. Y. Kurzweil-Segev, A.G. (Gutina), I. Popov, D. Golodnitsky, Y. Feldman, Phys. Chem. Chem. Phys. **18**, 10992 (2016)
30. Y. Miyazaki, T. Matsuo, H. Suga, J. Phys. Chem. B **104**, 8044 (2000). https://doi.org/10.1021/jp0007686
31. S. Khodadadi, A. Malkovskiy, A. Kisliuk, A.P. Sokolov, Biochim. Biophys. Acta (BBA) Proteins Proteom. **1804**, 15 (2014). https://doi.org/10.1016/j.bbapap.2009.05.006
32. J. Swenson, S. Cerveny, J. Phys. Condens. Matter **27**, 033102 (2014). https://doi.org/10.1088/0953-8984/27/3/033102
33. R. Hofmeister, A. Yariv, A. Agranat, J. Cryst. Growth **131**, 486 (1993). https://doi.org/10.1016/0022-0248(93)90201-7
34. P. Ben Ishai, C. De Oliveira, Y. Ryabov, Y. Feldman, A. Agranat, Phys. Rev. B **70**, 13 (2004)

35. P. Ben Ishai, C. De Oliveira, Y. Ryabov, Y. Feldman, A. Agranat, J. Non-Crystall. Solids **351**, 33 (2005). https://doi.org/10.1016/j.jnoncrysol.2005.04.073
36. P. Ben Ishai, A. Agranat, Y. Feldman, Phys. Rev. B **73**, 10 (2006). https://doi.org/10.1103/PhysRevB.73.104104
37. M. Gałązka, P. Zieliński, P. Szklarz, G. Bator, Phase Transitions **80**, 745 (2007). https://doi.org/10.1080/01411590701340300

Dielectric Thermodynamics of Organic Semiconductors

Jacopo Parravicini, Adele Sassella, and GianBattista Parravicini

Abstract We here consider the thermodynamic macroscopic behavior of organic molecular semiconductors (OMS) from the point of view of their dielectric response and phase trantitions. We focus on a couple of largely investigated crystals, namely α-quaterthiophene (4T) and tetracene (TEN). In such systems the evaluation of dielectric thermodynamics as a function of temperature is able not only to highlight the specific structure and patterns of phase transitions in markedly polymorphic compounds, but also to provide a practical estimation of the stability of each identified phase. The analysis carried out on such OMS can be considered a useful paradigm to apply the same procedure, interpretations, and concepts on the large and growing domain of organic materials for electronics.

1 Brief Landscape

Organic molecules form a very large set of compounds made of carbon and hydrogen atoms. Many of these compounds also contain other different elements called heteroatoms. Carbon atoms may present hybridization of the sp^2 orbitals producing the π molecular orbitals, when the electrons are completely delocalized over the entire molecule. The specific features of these organic molecules, called

J. Parravicini (✉)
Dipartimento di Fisica & Astronomia, Università degli Studi di Firenze, Sesto F., Firenze, Italy
e-mail: jacopo.parravicini@unifi.it

A. Sassella
Dipartimento di Scienza dei Materiali, Università di Milano-Bicocca, Milano, Italy
e-mail: adele.sassella@unimib.it

GB. Parravicini
Dipartimento di Fisica, Università degli Studi di Pavia, Pavia, Italy
e-mail: gianbattista.parravicini@unipv.it

J. Parravicini (ed.), *Fröhlich Entropy Estimation and Dielectric Calorimetry*, Advances in Dielectrics, https://doi.org/10.1007/978-3-032-21454-6_8

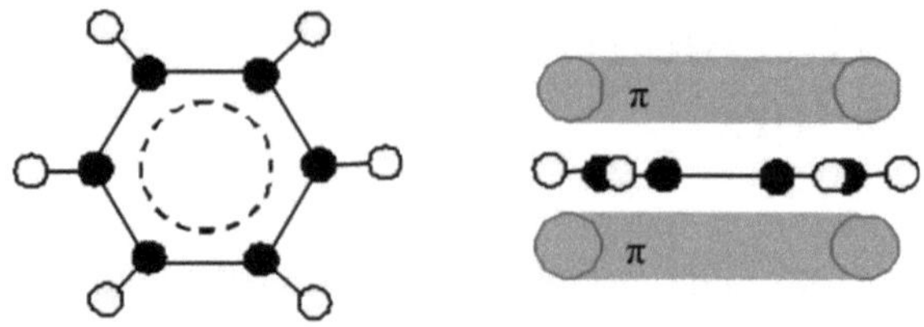

Fig. 1 Example of the structural formula of benzene (left, with black *C* atoms and white *H* atoms) and side view of the same molecule, with a sketch of the ring-shaped molecular π orbital (right). Reprinted with permission and adapted from [2], Fig. 1. (Courtesy of the original publisher. ©)

"conjugated" molecules[1] are mainly due to the presence of such π orbitals. It is worth recalling, e.g., that, in these compounds, the charge distribution within the π orbitals is the key factor affecting the separation of energy levels between LUMO and HOMO states[2] [1, 2]. These organic molecules can interact to form solid materials, notably molecular crystals with a crystalline structure, whose physical properties are mainly caused by the arrangement of the electrons lying in such π orbitals (Fig. 1) [3]. The key factor is the interactions and overlap of the π-electrons belonging to different molecules, which is due to the reciprocal configurations, such as distance or orientation, of their respective molecules. Actually, the typical solid state features of (organic) molecular solids or crystals still remain strongly correlated with the original features of the basic single molecules, since the inter molecular interactions are rather weak, typically van der Waals forces. This make such solids distinct from those formed through interaction of atoms [4]. A specific line of research on organic molecular materials concerns the semiconducting behavior that is exhibited by many of them, which makes them suitable to be employed in the vast domain of electronics and active materials [5–7]. This specific behavior represents a powerful boost to the investigations on these compounds, because their application properties are strongly determined by the specific basic molecule and by the solid structure that it generates, which can be amorphous, polycrystalline, or crystalline, and very often it gives rise to different possible polymorphs. For example, Fig. 2 shows different conjugated molecules which in solid state arrangement give rise to semiconducting behavior.

Similarly to conventional inorganic semiconductors, single crystals are the physical systems in which the study of the intrinsic features of the material can be more easily predicted and investigated, owing to the regularity of their underlying structures. This enables the achievement of reproducible, modellable, and controlled properties. Specifically, the high charge mobility displayed by π-conjugated organic molecular crystals makes them especially promising for electronic and

---

[1] As the property of possessing delocalized π-electrons is called conjugation.

[2] As usual, LUMO and HOMO mean "lowest unoccupied molecular orbital" and "highest occupied molecular orbital" respectively.

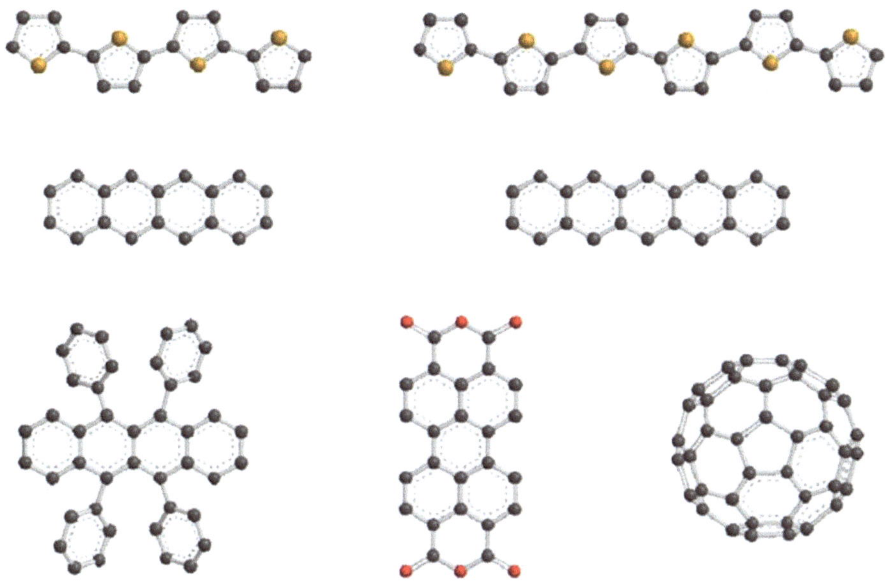

Fig. 2 Examples of the structural formula of some conjugated organic molecules leading, in the solid state, to organic semiconductors. First row: α-quaterthiophene (4T) and α-sexithiophene (6T); second row: tetracene (TEN) and pentacene (PEN); third row: rubrene (RUB), perylene-tetracarboxylic-dianhydride (PTCDA), and fullerene (C_{60}). *C* atoms are grey, *S* atoms yellow, and *O* atoms red; *H* atoms are not shown. Reprinted with permission and adapted from [2], Fig. 2. (Courtesy of the original publisher. ©)

optoelectronic applications. However, the above-recalled polymorphism of such compounds makes detailed studies on the effect of temperature and thermal cycles essential. This class of studies is a key element to obtain, from these materials, devices with reliable performances for semiconducting applications [8, 9]. Indeed, even when highly stable molecules are considered, the effects of temperature have been widely recognized as detrimental to the performance of devices employing such semiconductors. In this context, the role played by the physical properties of the active material remains the subject of ongoing investigations. More generally, the temperature effects on such materials, whose polymorphism involves phase transitions influencing their macroscopic properties, is a rich research topic. The onset of phase transformations are usually combined with the increasing presence of instabilities in the crystal properties due to a progressive increase of the molecular disorder. These phenomena can anticipate the actual structural transition by tens of degrees. Owing to the difficulty of their detection, these pre-transitional effects are often neglected, even though they are essential to identify the boundaries and the stability limits of each phase.

The overall picture discussed above indicates that both temperature-resolved thermodynamic investigation and dielectric measurements can provide very useful tools for addressing several still-open issues about these compounds. The combination of such approaches is particularly effective in tackling some open questions, especially regarding the amplitude of phases, the corresponding order degree, the features of phase transitions, the dependence on thermodynamic cycles, the presence of hysteretic effects, etc. [10, 11]. Notably, the application of Fröhlich entropy evaluation technique [12] has been demonstrated to be fruitful in the analysis of the specific features of some organic semiconducting compounds among the most widely investigated, i.e. α-quaterthiophene (4T) and tetracene (TEN) [10, 11].

2 Investigation of Organic Crystalline Molecular Semiconductors

Semiconducting organic molecular crystals represent a large class of materials that have attracted considerable interest over the past several years, owing to their electrical properties which make them very interesting for the development of novel electronic devices [13–16]. In this wide set of rubrenes, oligoacenes (e.g. tetracene and pentacene), oligothiophenes (e.g. quaterthiophene and sexithiophene), and pyridines are among the most interesting and studied. On one hand, from an application point of view, they have a large potential to realize organic electronic devices, such as transistors, solar cells, sensors. On the other hand, they host several effects in the field of condensed-matter physics, in particular polymorphisms, different crystalline phases, closely spaced in temperature and energy, and complex transition phenomena [17–20]. This multifaceted behavior poses a significant experimental challenge for achieving a trustworthy description of the features of each compound depending on temperature and thermodynamic history. Such conditions need the use of several different experimental approaches.

In this vein, a fruitful experimental approach is the estimation of Fröhlich entropy (FE). Such physical quantity is the entropy variation of a dielectric due to the application of a static electric field, which is interpreted as a reliable figure of merit of the order and disorder degree[3] [12, 21–24]. This is the reason why the use of FE estimation is a powerful tool to highlight some unclear features in this class of materials. Particularly, FE estimation is employed in a number of studies on phases and phase transitions of two among the most relevant organic molecular semiconductors, namely α-quater-thiophene (4T, Fig. 3) and tetracene (TEN) crystals [10, 11], in single-crystal configuration (Fig. 3). Although these materials were investigated through several experimental techniques, e.g. X-ray and neutron diffraction, optical spectroscopy (fluorescence, reflection, ellipsometry), and Raman spectroscopy, nevertheless several issues have nevertheless remained only partially understood [15, 19, 25–32]. Actually, the response of 4T and TEN crystals, in terms of their FE as a function of temperature, reveals interesting and peculiar features.

[3] See Sect. 3 in chapter "Introduction: Dielectric Calorimetry and Fröhlich Entropy Estimation".

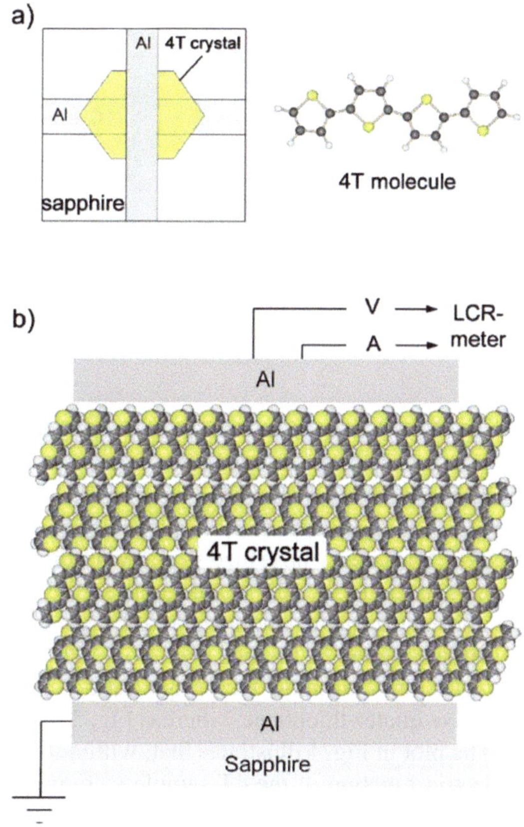

Fig. 3 Arrangement of the α-quater-thiophene sample investigated (from [10]). (**a**)-left: top-view of the experimental disposition of the sample; (001)-oriented 4T single crystal (grayish) sandwiched between two Al stripes to form capacitor plates on a sapphire substrate. (**a**)-right: scheme of the chemical structure of a generic 4T molecule (C-dark gray, H-white, S-pale gray). (**b**) Sketch of the sample cross section, showing the molecular arrangement in the 4T crystal. Reprinted with permission and adapted from [10], Fig. 1. (Courtesy of the original publisher. ©)

2.1 Phase Transitions

Organic crystalline semiconductors (OCS) generally experience phase transitions in response to temperature changes [28–30]. Their evolution exhibits specific characteristics, such as the fact that phase transitions usually coincide with the occurrence of instabilities. These are brought about by a slow increase in molecular disorder and have macroscopic implications. This behavior can be observed in temperature intervals as wide as tens of degrees straddling the specific transition [33] and exhibits characteristics similar to pretransitional (or premelting) effects, as seen, for example, in the previously discussed Ga nanoparticles (chapter "Fröhlich Entropy of Nanoparticles: Investigation of Phases, Phase Transitions and Melting"). Indeed, these effects can be observed in temperature ranges spanning several degrees around the precise transition [33].

The pretransitional phenomena in OCS are often overlooked due to their elusive nature. Nevertheless, when the critical temperatures and the associated instability of phases must be determined, it is necessary to consider them. The properties of transitions are expected to be associated with the size of the investigated system,

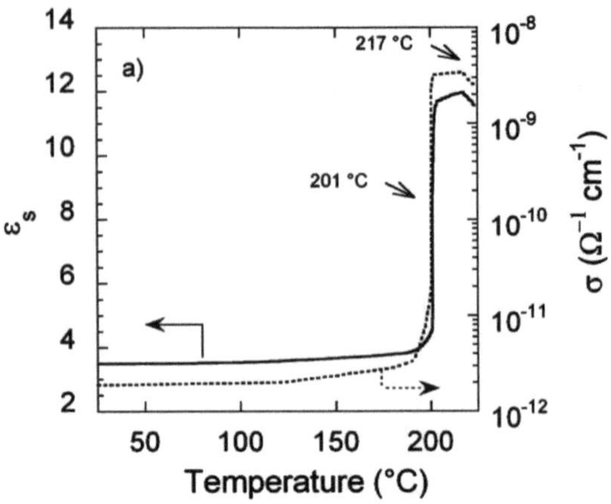

Fig. 4 4T thin-film sample: measured dielectric response ε_s and corresponding conductivity σ as a function of temperature T. Reprinted with permission and adapted from [10], Fig. 2. (Courtesy of the original publisher. ©)

specifically when low-dimensional or micro- (and nano-)metric samples (thin films, powders, etc.) are taken into consideration. This is comparable to what occurs in metallic particles, whose size strongly modifies the physical features of phases and transitions[4] [21, 34, 35]. In this context, FE assessment can be profitably used. If we consider α-quater-thiophene (4T), we may see that the behavior of FE is able to describe in detail the thermal evolution of the structural transitions in the 4T crystal. Namely, Fig. 4 displays the temperature dependence of the permittivity of single crystal α-quater-thiophene, whereas Fig. 5 displays the corresponding FE.

The plot in Fig. 4 illustrates that, within the temperature range of 25–190 °C, the dielectric function of the 4T sample ε_s remains virtually constant, exhibiting only a slight increase with temperature. At approximately 190 °C, ε_s rises more rapidly with temperature, and at 201 °C, it exhibits a step. Above 201 °C, the curve still has a positive slope until the melting temperature $T_m = 217\,°C$ is reached; for $T > T_m$, the slope of the $\varepsilon_s(T)$ curve becomes negative. The increase of S_E at 190 °C is the first indication of transition phenomena.

When Fröhlich entropy is taken into account (Fig. 5), the crucial transition temperature is indicated by the sharp peak at $T = 201\,°C$. Then, there is a T region where S_E stays positive and nearly constant, up to 217 °C. This indicates the presence of two separate ordered-solid stable phases with marginally differing levels of order, which undergo a structural change. We find that this transition begins at 190 °C and concludes at 204 °C, with a temperature extent of roughly 15 °C.

Moreover, when the temperature reaches $T_m = 217\,°C$, FE exhibits a step-like discontinuity, without notable pretransitional effects, as the entropy shifts from positive to negative. According to Fröhlich's interpretation elsewhere recalled (Sect. 3.3 in chapter the "Introduction: Dielectric Calorimetry and Fröhlich Entropy Estimation" and Sect. 1 in chapter "Meaning and Applications of Fröhlich Entropy"), this

[4] See chapter "Fröhlich Entropy of Nanoparticles: Investigation of Phases, Phase Transitions and Melting".

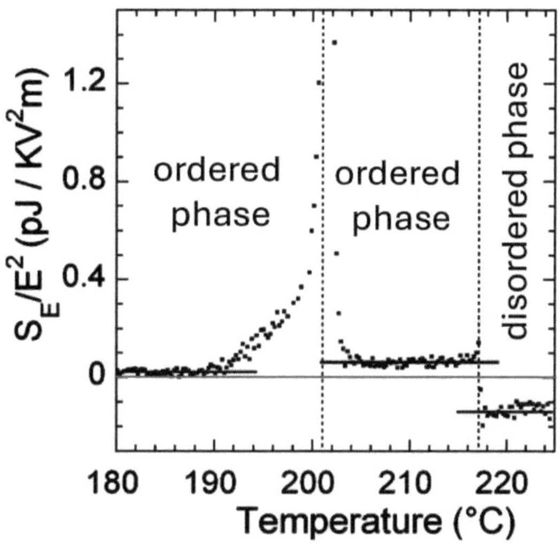

Fig. 5 4Y thin-film sample: Fröhlich entropy for unitary field, S_E/E^2, as a function of T, obtained from the derivative of ε_s in the temperature region 180–225 °C (correlated to the $\varepsilon_s(T)$ displayed in Fig. 4) where the order-order (solid-solid-like) and (order-disorder) (solid-liquid-like) transitions are found. The horizontal gray line indicates the 0-value of FE. The transition critical temperatures are indicated by the vertical dotted lines. The first two ordered, solid phases, regions (positive FE) are highlighted, together with the last disordered, liquid phase, region (negative FE). Reprinted with permission and adapted from [10], Fig. 2. (Courtesy of the original publisher. ©)

pattern indicates a change from an ordered solid-like phase ($S_E > 0$) to a disordered liquid-like one ($S_E < 0$). Lastly, FE has a constant negative value for $T > T_m$, which is indicative of a stable liquid-like system. It is important to note that the induced S_E value computed for 4T in its liquid phase is comparable to the value already found in other organic liquids for which FE has been evaluated [36].

In summary, from the above-discussed paradigmatic material, we may reasonably generalize and deduce that the FE evaluation technique in OCS is suitable for providing a measure of the state-of-order of solid and liquid states as well as a tool to identify the nature and evolution of temperature-induced phase transitions together with their temperature extent [12].

2.2 *Stability of Phases*

The evaluation of the stability degree of a certain phase of solid-state matter, here an organic crystalline semiconductor, has been addressed in several physical systems, where standard experimental approaches are employed. These usually exploit global thermodynamic quantities such as enthalpy, entropy, Gibbs free energy, and similar functions [37–45], which are the basis for a variety of experimental methods.

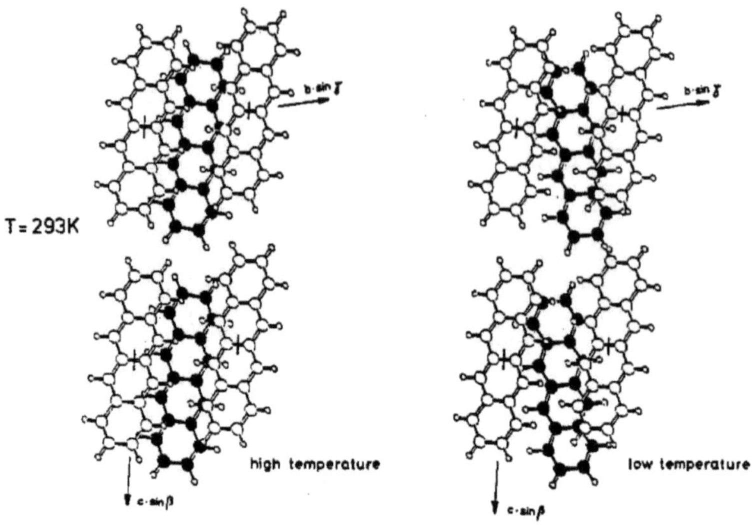

Fig. 6 Arrangement of tetracene (TEN) molecules viewed along the [100] direction of the high-temperature (HT) and low-temperature (LT) phases. Reprinted with permission and adapted from [19], Fig. 9. (Courtesy of the original publisher. ©)

Fröhlich entropy belongs to the set of such useful thermodynamic potentials; thus, along the same line, the FE evaluation technique can be a valuable option for investigating the stability degree of a material.[5] More specifically, organic molecular crystals can be considered a prototypical class of systems whose phase stability is often an outstanding question.

Among what is present in the literature, FE was exploited in tetracene (TEN) organic crystals, which is one of the most interesting OMS due to its applications in field-effect transistors and light-emitting diodes and its use as a dye in photonic and luminescence techniques [46–48]. These applications account for the large interest about such compound, which displays a quite complex structure (see, e.g., Figs. 6 and 7) [19, 49]. In OMS, and especially in TEN, the coexistence of several phases across wide temperature intervals had previously been observed, where a complex polymorphism in relation to thermodynamic parameters raises the question of carrying out an exhaustive investigation of phase stability [11]. Indeed, the polymorphic features of TEN were experimentally demonstrated through a variety of experimental methods, including optical and Raman spectroscopies, X-ray and neutron diffraction, and others [15, 19, 25–27, 31, 32, 50]. Nonetheless, several issues still remain open, which FE estimation, in comparison with the already acquired TEM literature data, was demonstrated to be able to illuminate [11]. We

[5] See chapters "Introduction: Dielectric Calorimetry and Fröhlich Entropy Estimation" and "Meaning and Applications of Fröhlich Entropy", in particular Sect. 2.3 in chapter "Meaning and Applications of Fröhlich Entropy".

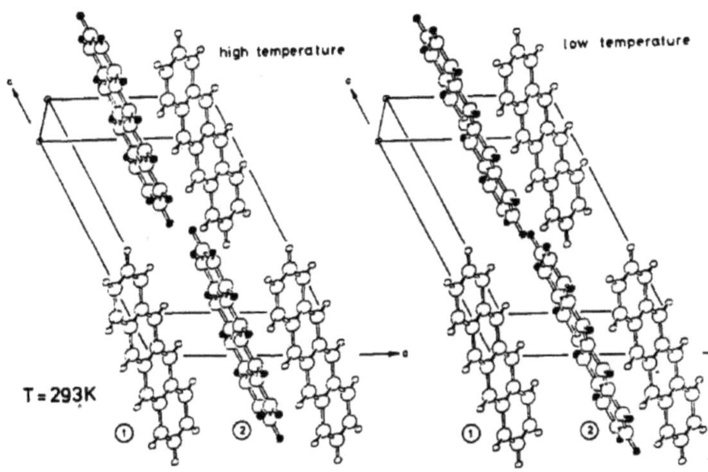

Fig. 7 Molecular arrangement within the unit cell of the high-temperature (HT) and low-temperature (LT) structures of crystalline tetracene (TEN). Reprinted with permission and adapted from [19], Fig. 8. (Courtesy of the original publisher. ©)

also note that this large experimental effort makes this organic crystal a test system, paradigmatic for several analogous organic crystals.

The use of FE evaluation to investigate the thermodynamic stability of TEN crystals is accomplished by starting from the measurement of the dielectric response over a wide temperature range (12–295 K). The results are shown in Figs. 8a and 9a for the cooling and heating ramps, respectively. Three permittivity peaks at 50, 153, and 223 K are displayed in Fig. 8a (cooling), indicating three transitions; a significantly perturbed T region is located between 125 and 150 K. At the same temperatures, the corresponding FE map (Fig. 8b) shows a transition pattern, i.e. sharp sign changes, steps, and oscillations.

Three roughly stable phases, with different degrees of order, are found below and just above the region of instability, and they are pinpointed by a nearly flat FE trend. The continuous increase of the FE positive values with temperature, which can be observed above 223 K, suggests a noticeable increase in molecular correlations. On the other hand, the heating stage exhibits a markedly different behavior (Fig. 9). The observed peaks of $\varepsilon_r(T)$ (Fig. 9a) are significantly sharper and higher, and there is no a disturbed temperature zone in between the peaks. In contrast, the behavior of FE pinpoints six phases by highlighting five distinct phase transitions (Fig. 9b). The three phases at the lowest temperatures (25–45 K, 45–105 K, and 105–144 K) exhibit FE with a positive sign and a virtually flat trend, indicating global stability. Conversely, a marked change in the order state is observed in the high temperature phases (144–178 K, 178–207 K, and 207–280 K). Specifically, the phase at the highest temperature still exhibits negative FE values, suggesting a very disordered state. The two phases identified between 105 and 178 K overlap, therefore revealing

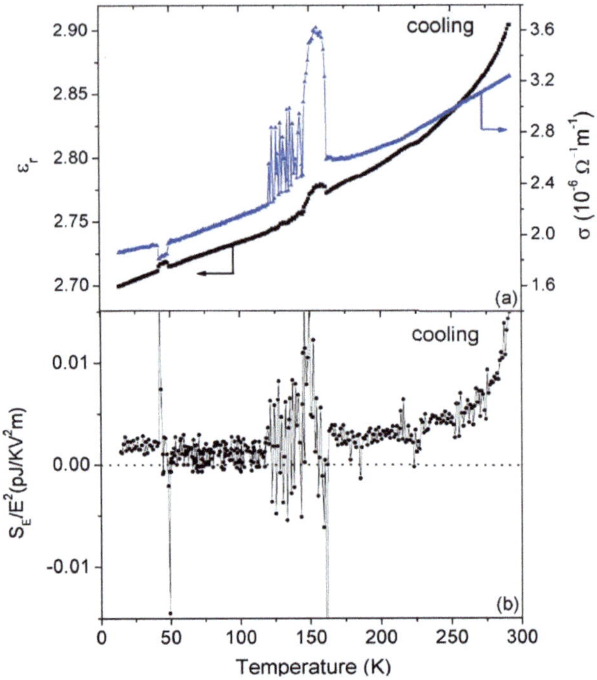

Fig. 8 (**a**) relative dielectric constant ε_r and of the conductivity σ vs. T during cooling from RT down to 12 K of a 3-μm-thick TEN single crystal. (**b**) $s = s(T)$, obtained from the derivative of the dielectric constant of Plot (**a**). We note that the instability in the 125–160 K region is due to the overlapping of two well-distinct transitions, which are highlighted in the heating stroke reported in Plot 9. In both Plots the critical temperatures are 50, 153, and 223 K. Reprinted with permission and adapted from [11], Fig. 1. (Courtesy of the original publisher. ©)

the meaning of the instability region observed during the cooling stage (Fig. 8b) [11].

It is finally worth noting the highly "noisy" trend of FE, with rapid oscillations and several sudden changes of sign (Fig. 8b). This behavior highlights the temperature region where marked hysteresis phenomena are observed, as demonstrated by the measurements in the heating stage (Fig. 9), and identifies the presence of a typical out-of-equilibrium (non-ergodic) occurrences [11]. Such an evident non-ergodic response is indicated by the strong dependence on the thermodynamical history of the sample and it can be evidenced both in the change of critical temperatures of the identified transitions and in the change of the state-of-order of each phase [11, 51].

3 Conclusions

The dielectric thermodynamic approach for the investigation of organic molecular crystals, which is based on Fröhlich's theory and his interpretation of the induced entropy, is demonstrated to provide a global landscape of the phases and state-of-order of such materials. In particular, the Fröhlich entropy evaluation technique is a trustworthy experimental method allowing the extraction of a large amount of

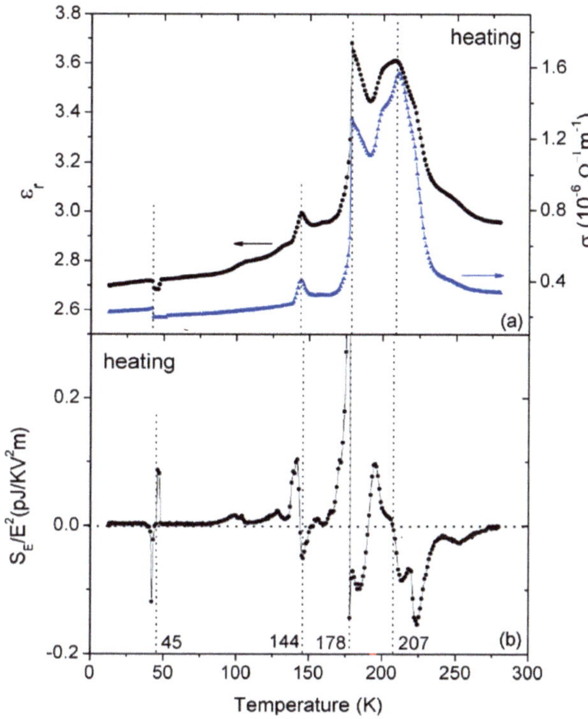

Fig. 9 (**a**) relative dielectric constant ε_r and of the conductivity σ vs. T during heating from 12 K up to RT of the same sample as in Fig. 8 and with the same temperature scan rate. (**b**) $s = s(T)$, obtained from the derivative of the dielectric constant in Plot (**a**). In both Plots the critical temperatures are 45, 105, 144, 178, and 207 K. Reprinted with permission and adapted from [11], Fig. 2. (Courtesy of the original publisher. ©)

information, which we may summarize in the following items:

1. detailed mapping of the pattern of the different phase transitions as a function of temperature;
2. identification of the critical temperatures and the correlated pre- and post-transition features.
3. estimation of the stability level of the explored phases;
4. continuous monitoring of the evolution of the state of order in the sample.

In summary, the evaluation of dielectric thermodynamic parameters, especially the Fröhlich entropy, allows one to highlight the nature and evolution of temperature-induced phase transitions in OMS, and to obtain information on the features displayed during the transition, as well as before and after it. Through FE evaluation, the associated critical temperatures are identified in detail, together with the temperature extent of the stability of the different crystalline phases, observing and studying pre- and post- transitional effects, as well as premelting ones.

References

1. A. Sassella, M. Campione, A. Borghesi, La Rivista del Nuovo Cimento **31**, 457 (2008)
2. A. Sassella, Cryst. Res. Technol. **48**, 840 (2013)
3. A. Davydov, *Theory of Solids* (Nauka, Moscow, USSR, 1976)
4. A. Davydov, *Theory of Molecular Excitons* (McGraw-Hill, New York, 1962)
5. R. Farchioni, G. Grosso, *Organic Electronic Materials* (Springer, Heidelberg, 2001)
6. M. Schwörer, H.C. Wolf, *Organic Molecular Solids* (Wiley VCH, Weinheim, 2005)
7. W. Brütting (ed.), *Physics of Organic Semiconductors* (Wiley VCH, Weinheim, 2005)
8. Z. Rang, A. Haraldsson, D. Kim, P. Ruden, M. Nathan, R. Chesterfield, C. Frisbie, Appl. Phys. Lett. **79**, 2731 (2001)
9. H. Najafov, B. Lee, Q. Zhou, L. Feldman, V. Podzorov, Nat. Mater. **9**, 938 (2010)
10. A. Sassella, D. Braga, M. Campione, T. Ciabattoni, M. Moret, J. Parravicini, GB. Parravicini, J. Appl. Phys. **109**, 013529 (2011)
11. GB. Parravicini, M. Campione, F. Marabelli, M. Moret, A. Sassella, Phys. Rev. B **86**, 024107 (2012)
12. J. Parravicini, G. Parravicini, Res. Phys. **28**, 104571 (2021)
13. R.W. de Boer, T. Klapwijk, A. Morpurgo, Appl. Phys. Lett. **83**, 4345 (2003)
14. S. Verlaak, D. Cheyns, M. Debucquoy, V. Arkhipov, P. Heremans, Appl. Phys. Lett. **85**, 2405 (2004)
15. A. Pivovar, J. Curtis, J. Leao, R. Chesterfield, C. Frisbie, Chem. Phys. **325**, 138 (2006)
16. V. Podzorov, E. Menard, A. Borissov, V. Kiryukhin, J. Rogers, M. Gershenson, Phys. Rev. Lett. **93**, 086602 (2004)
17. D. Gundlach, T. Jackson, D. Schlom, S. Nelson, Appl. Phys. Lett. **74**, 3302 (1999)
18. T. Siegrist, C. Kloc, J. Schön, B. Batlogg, R. Haddon, S. Berg, G. Thomas, Angew. Chem. Int. Ed. Engl. **40**, 1732 (2001)
19. U. Sondermann, A. Kutoglu, H. Bässler, J. Phys. Chem. **89**, 1735 (1985)
20. R.G. Della Valle, E. Venuti, A. Brillante, A. Girlando, J. Phys. Chem. A **110**, 10858 (2006)
21. GB. Parravicini, A. Stella, M. Ungureanu, P. Merli, A. Migliori, R. Kofman, Phys. Status Solidi B **237**, 374 (2003)
22. J. Parravicini, E. DelRe, A. Agranat, GB. Parravicini, Phys. Rev. B **93**, 094203 (2016)
23. J. Parravicini, E. DelRe, A.J. Agranat, GB. Parravicini, Nanoscale **9**, 9572 (2017)
24. J. Parravicini, Phys. B **541**, 54 (2018)
25. A.F. Prikhotko, A.F. Skorobogatko, Opt. Spectrosc. **20**, 33 (1996)
26. G. Vaubel, H. Baessler, Mol. Cryst. Liq. Cryst. **12**, 39 (1970)
27. D. D. Kolendriski, M. V. Kurik, Yu. P. Piriatinskii, Phys. Status Solidi B **91**, 741 (1979)
28. A. Sassella, M. Campione, A. Papagni, C. Goletti, G. Bussetti, P. Chiaradia, V. Marcon, G. Raos, Chem. Phys. **325**, 193 (2006)
29. S. Tavazzi, M. Campione, Appl. Phys. Lett. **88**, 071918 (2006)
30. M. Campione, S. Tavazzi, M. Moret, W. Porzio, J. Appl. Phys. **101**, 083512 (2007)
31. R. Jankowiak, J. Kalinowski, M. Konys, J. Buchert, Chem. Phys. Lett. **56**, 549 (1979)
32. E. Venuti, R.G. Della Valle, L. Farina, A. Brillante, M. Masino, A. Girlando, Phys. Rev. B **70**, 104106 (2004)
33. GB. Parravicini, A. Stella, P. Tognini, P. Merli, A. Migliori, P. Cheyssac, R. Kofman, Appl. Phys. Lett. **82**, 1461 (2003)
34. GB. Parravicini, A. Stella, M. Ungureanu, R. Kofman, Appl. Phys. Lett. **85**, 302 (2004)
35. GB. Parravicini, A. Stella, M. Ungureanu, P. Merli, A. Migliori, P. Cheyssac, R. Kofman, Eur. Phys. J. D **24**, 219 (2003)
36. J. Jadzẏn, G. Czechowski, J. Phys. Chem. B **111**, 3727 (2007)
37. L.D. Landau, E.M. Lifshitz, *Electrodynamics of Continuous Media*, 2nd edn. (Pergamon Press, Oxford, UK, 1960)
38. L. Landau, E. Lifshitz, *Statistical Physics*, 3rd edn. (Pergamon Press, Oxford, UK, 1980)

39. H.B. Callen, *Thermodynamics and an Introduction to Thermostatistics*, 2nd edn. (John Wiley & Sons, New York, 1985)
40. A.I. Kitajgorodskij, Acta Cryst. **18**, 585 (1965)
41. D. Miracle, O. Senkov, Acta Mater. **122**, 448 (2017)
42. J. Muscat, V. Swamy, N. Harrison, Phys. Rev. B **65**, 224112 (2002)
43. S. Guo, C. Liu, Pro. Nat. Sci. Mater. **21**, 433 (2011)
44. S. Tessier, J. Brennecke, M. Stadtherr, Chem. Eng. Sci. **55**, 1785 (2000)
45. J.P. Sethna, *Statistical Mechanics: Entropy, Order Parameters, and Complexity*, 1st edn. (Oxford University Press, Oxford, UK, 2006)
46. M. Schwoerer, H. Wolf, *Organic Molecular Solids* (Wiley-VCH, Weinheim, 2006)
47. C. Adachi, T. Tsutsui, Fundamentals of luminescence: Luminescence of organic compounds, in *Fundamentals of Phosphors* (CRC Press - Taylor & Francis, New York, 2006), pp. 51–59
48. H. Dong, C. Wang, W. Hu, Chem. Commun. **46**, 5211–5222 (2010)
49. J. Robertson, V. Sinclair, J. Trotter, Acta Crystallogr. **14**, 697–704 (1961)
50. J. Glinski, J. Kalinowski, S. Stizza, R. Bernardini, A. Bianconi, Solid State Commun. **38**, 155 (1981)
51. A. Bokov, Z. Ye, J. Mater. Sci. **41**, 31 (2006)

Order–Disorder Transitions of Ferroelectric Perovskites by Fröhlich Entropy Investigations

Peng Tan, Yuwen Qiao, Zijian Zhang, and Hao Tian

Abstract The order-disorder phase transition, a crucial structural phase transition, has attracted significant attention in ferroelectric perovskites research. During the phase transition, the dynamic evolution of the ordered-disordered states of polarization frequently yields novel physical effects and serves as the important origin for excellent macroscopic properties. The Fröhlich entropy offers an effective way of exploring the order-disorder phase transition in perovskite ferroelectrics. It can reveal the responses of polarization units under the action of an electric field during the phase transition and indicate the changes in the degree of polarization order. In this chapter, we first provide a brief overview of the order-disorder phase transition in perovskite ferroelectrics. Then, the intrinsic correlation mechanism between Fröhlich entropy and the phase transition is elaborated. We also introduce the key roles of Fröhlich entropy in revealing the structural evolution and the origin of properties during the phase transition, demonstrating its value in understanding the physical essence of materials. We hope that the chapter can assist readers in comprehensively understanding the relevant schemes for investigating ferroelectric phase transition by Fröhlich entropy, and provide useful references for related research.

1 Order–Disorder Phase Transitions in Ferroelectric Perovskites

Ferroelectrics refer to crystals that possess spontaneous polarization within a certain temperature range, and their spontaneous polarization can be reversed or re-oriented by an external electric field. Ferroelectrics with a perovskite-structured lattice are called ferroelectric perovskites.

P. Tan · Y. Qiao · Z. Zhang · H. Tian (✉)
School of Physics, Harbin Institute of Technology, Harbin, China
e-mail: tanpeng@hit.edu.cn; tianhao@hit.edu.cn

J. Parravicini (ed.), *Fröhlich Entropy Estimation and Dielectric Calorimetry*, Advances in Dielectrics, https://doi.org/10.1007/978-3-032-21454-6_9

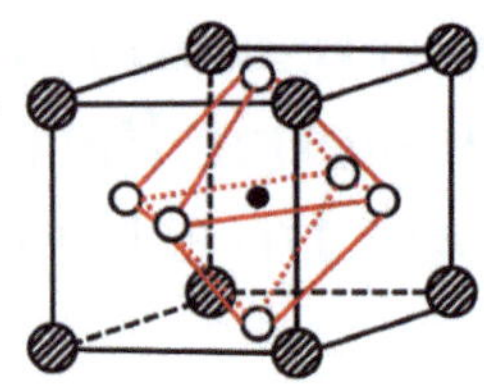

Fig. 1 ABO_3 structure

Most perovskite ferroelectrics can be represented by the general formula ABO_3, and their perovskite structures can be presented with a simple cubic lattice, as shown in Fig. 1. In this structure, the A-site ions are located at the vertices. These ions are usually rare-earth or alkali metal elements with a large ionic radius, and a coordination number of 12. The B-site ions are at the body centre, and are generally elements with a relatively small ionic radius, such as transition metal elements, with a coordination number of 6. The O ions are located at the six face centres. These O ions form an octahedral structure, and the B-site ions are located at the centre of the octahedron. Overall, the crystal can be regarded as being composed of octahedra connected by shared vertices, and the voids formed after the connection of each octahedron are occupied by A-site ions.

The spontaneous polarization of perovskite ferroelectrics mainly stems from the movement of B-site ions away from the centre of the octahedron. Because the displacement of B-site ions from the centre usually occurs along one of the high-symmetry directions of the octahedron, the direction of spontaneous polarization also typically follows these directions. Perovskite ferroelectrics combine the characteristics of perovskite structure and ferroelectricity, demonstrating their unique physical properties and diverse functions. Their excellent ferroelectric, piezoelectric, dielectric, photovoltaic, and catalytic properties endow them with significant practical value in modern technology, providing great convenience and benefits to human life and industrial production.

When external constraints such as temperature and pressure change continuously, under specific conditions (temperature, pressure, and other constraints reach certain specific values), ferroelectrics change from one structural state to another, a phenomenon called ferroelectric phase transition. During a ferroelectric phase transition, the ferroelectric material undergoes a phase structure change without physical state change. The ferroelectric phase transition is a solid-to-solid process, but the crystal structure changes significantly. Therefore, this type of phase transition is called structural phase transition. Landau pointed out that phase transitions are always accompanied by changes in structural symmetry and degree of order [1]. According to the phase transition mechanism, ferroelectric phase transitions can be divided into displacive and order–disorder types [2], as shown in Fig. 2. In the displacive-type ferroelectric phase transition, the phase transition can be associated with structural distortion, which promotes the relative cationic displacement and allows a change in spontaneous polarization. In the order–disorder phase transition, the spontaneous polarization is controlled by the ordered distribution of atoms or atomic groups. In fact, many observed ferroelectric phase transitions possess

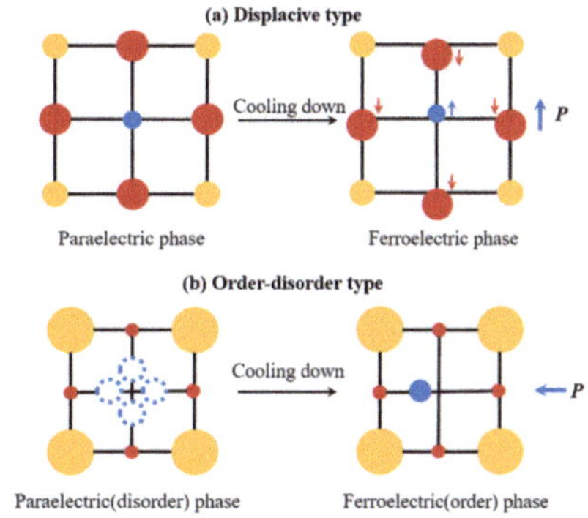

Fig. 2 Schematic illustrations of the (**a**) displacive-type and (**b**) order–disorder-type phase transitions in ferroelectric perovskites

characteristics of both types. In this book, we focus on introducing the order–disorder component in ferroelectric phase transitions.

In some ferroelectrics, under specific constraint conditions, certain atoms or atomic groups have two or more equilibrium positions and can transition between them. This leads to a disordered distribution of atoms and atomic groups at these positions. When the constraint conditions are changed, these atoms or atomic groups preferentially occupy a specific equilibrium position; thus, their distribution becomes ordered. This is called an order–disorder transition, which can be described by the transverse-field Ising model. The order–disorder component is present in the vast majority of ferroelectric phase transitions. With the development of microstructure characterisation techniques, many ferroelectric phase transitions that were once thought to be purely displacive have shown certain order–disorder characteristics. For example, $BaTiO_3$ has traditionally been regarded as a typical displacive ferroelectric. However, subsequent studies have shown that, above the Curie temperature (T_C), the Ti ions at the B-site are distributed in a disorderly manner among the eight equilibrium positions along $<111>_C$. During phase transition, the disordered state transforms into an ordered state. On one hand, the Ti ions are preferentially distributed among the four positions. On the other hand, static displacements occur along $[001]_C$ owing to the condensation of the soft mode [3]. As a typical perovskite ferroelectric, the phase transition of $PbTiO_3$ also exhibits order–disorder characteristics. Above T_C, the Pb ions are distributed in a disorderly manner among the six equilibrium positions along the $<100>_C$ direction. Only during the phase transition do they preferentially occupy one of these positions, resulting in a static displacement along $[001]_C$ [4].

In perovskite ferroelectrics, the characteristics of local polarization distribution, which are caused by the lattice structure and ion arrangement, are the core factors that influence the degree of order. These characteristics determine the transition between the ordered and disordered states. Moreover, the polarization characteristics of perovskite ferroelectrics are important sources of their numerous functions. The excellent properties of the material near the phase transition temperature are closely related to the ordered or disordered characteristics of local polarization. Therefore, studying the order-disorder phase transition in perovskite ferroelectrics or the order–disorder component in ferroelectric phase transitions is of great significance and value in exploring the microscopic mechanism of ferroelectricity in perovskite ferroelectrics, enriching the phase transition theory, guiding performance optimisation, and designing new functional materials.

2 Fröhlich Entropy and Order–Disorder Transition

The thermodynamic theory is important in studying phase transition processes. According to thermodynamic theory, after appropriately selecting independent variables, one thermodynamic function can completely determine the equilibrium properties of a homogeneous system. This function is called the characteristic function. When considering the case in which the independent variables are temperature T, stress $\boldsymbol{\sigma}$, and electric field $\mathbf{E}$, the characteristic function is the Gibbs free energy G.[1] During a phase transition, if the derivatives of G up to order $(n-1)$ are continuous while the n-th derivative is discontinuous, it is called an n-th-order phase transition. Entropy, commonly denoted by S, is the first-order derivative of G; that is,

$$S = -\left(\frac{\partial G}{\partial T}\right)_{\mathbf{X},\mathbf{E}} \tag{1}$$

Generally, second- and higher-order phase transitions are referred to as continuous phase transitions. Thus, in a first-order phase transition, entropy S is discontinuous, whereas in a continuous phase transition, entropy S is continuous.

The concept of entropy S was proposed by Clausius in 1865. In thermodynamics, entropy S, similar to free energy, is an important parameter that characterizes the state of matter. Its physical meaning is a measure of the degree of order (disorder) of a system. The more disordered the system, the greater the entropy; conversely, the more ordered the system, the smaller the entropy. The entropy of a system is determined by its internal structure and the influence of the external environment. Entropy has become a key parameter reflecting the behavioural characteristics of materials and guiding the optimisation of material properties. For example, in popular materials involving multi-component, multi-phase structures, and complex microstructures, such as high-entropy alloys and high-entropy ceramics, researchers have found that by introducing multiple elements or groups to increase the configurational entropy of the materials, the complex lattice structures can be effectively stabilized, and the phase stability and properties of the materials can be improved [5, 6].

[1] See chapter "Elements of Thermodynamics of Dielectrics and Basic Theory of Fröhlich Entropy" and in particular Sect. 2.3.

In perovskite ferroelectrics, the polarization structure is often complex. Near the phase-transition range, with a change in the external constraint conditions, such as temperature, the polarization structure evolves dramatically. Such changes in the polarization structure are often accompanied by alterations in the order–disorder state. Entropy S provides an important approach for studying the changes in the order–disorder polarization state in perovskite ferroelectrics and serves as a powerful basis for revealing the dynamic processes in order–disorder phase transitions. During the phase transition process, considering the response characteristics of ferroelectric polarization under an applied electric field, Fröhlich pointed out that in a reversible isothermal system, the entropy change can be written as [7]

$$S(T, \mathbf{E}) = S_0(T) + S_{\mathbf{E}}(T, \mathbf{E}) \tag{2}$$

where T is the absolute temperature, $S_0(T)$ is the entropy at a zero field, and $S_{\mathbf{E}}(T)$ is the field-dependent part of entropy. Fröhlich et al. established the relationship between this entropy change and the dielectric function, that is:

$$S(T, \mathbf{E}) = S_0(T) + \frac{1}{2}\frac{\partial \varepsilon}{\partial T}\mathbf{E}^2 \tag{3}$$

This is called Fröhlich entropy, where $\varepsilon = \varepsilon_0 \varepsilon_s$, is a measure of the dielectric response, ε_0 is the vacuum permittivity, and ε_s is the static relative permittivity.

Dielectric response refers to the physical phenomenon caused by the slight displacement of charges in a material under an applied electric field. In perovskite ferroelectrics, the dielectric response mainly originates from three contributions: electronic polarization, ionic polarization, and orientation polarization of dipoles. It is often characterized by dielectric temperature spectrum tests and is one of the most common methods for studying phase transition processes. The larger the dielectric constant, the higher the degree of polarization response of the material in the electric field. Therefore, the dielectric response essentially reflects the polarization characteristics of the material. Fröhlich entropy links the entropy change under the action of an electric field, dielectric properties, and the order–disorder polarization state. According to this formula, Fröhlich entropy can be directly calculated from the dielectric response function of the ferroelectrics. Its change trend can provide entropy information on the temperature evolution of the material, making it highly suitable for the phase transition region, where there are drastic changes in the order (disorder) degree.

Assumptions are made in Fröhlich theory. That is, the volume of the material remains unchanged under an external electric field; the applied electric field only orientates the internal dipoles without stretching the dipoles, and the electrostriction effects can be neglected. The only independent variables are the temperature T and external electric field $\mathbf{E}$. Fortunately, in actual dielectric response tests, the generally applied electric field strength is several orders of magnitude smaller than the coercive field of the material. This ensures that the dynamic process conforms well to the Fröhlich thermodynamic relationship model. As a result, the Fröhlich entropy obtained from convenient dielectric tests has become a reliable way to measure the order–disorder state in materials with the advantage of intuitively reflecting the degree of polarization order/disorder.

In Formula (3), the part that does not contain the electric field in $S_E(T, E)$ is represented by $s(T)$, that is:

$$s(T) = \frac{\varepsilon_0}{2}\frac{\partial \varepsilon_s}{\partial T} \tag{4}$$

$s(T)$ is independent of the electric field strength and closely related to the dielectric response of the material. It can be used to illustrate the order-disorder state of the system. It shows that the application of an electric field increases the entropy if $\partial\varepsilon_s/\partial T > 0$ and decreases the entropy when $\partial\varepsilon_s/\partial T < 0$. The positive and negative values of $s(T)$ represent different degrees of order. When dipoles within a material display solid-like characteristics, forming an ordered system in a stable low-energy state, the external electric field causes some dipoles to undergo discontinuous quantum transitions. This intensifies the disorder of the system, leading to an increase in entropy. At this point, $\partial\varepsilon_s/\partial T$ is positive, signifying that $s(T) > 0$. Conversely, consider a physically disordered system with electric dipoles. The external electric field can reorient the electric dipoles, which were originally arranged randomly, thus increasing order in the system. As a result, the entropy change caused by the electric field is negative, $\partial\varepsilon_s/\partial T < 0$, corresponding to $s(T) < 0$. In dielectric response tests, because the applied electric field strength is far less than the coercive field strength, the external electric field and electric polarization intensity satisfy a linear relationship. Thus, $\partial\varepsilon_s/\partial T > 0$ implies that the system is in an ordered state (dipolar solid-like state), while $\partial\varepsilon_s/\partial T < 0$ indicates a disordered state (dipolar liquid-like state). When phase transformations are considered, an order/order (disorder/disorder) transition maintains the positive (negative) sign of $s(T)$ with a possible variation of its magnitude, whereas an order/disorder (or disorder/order) is signaled by a variation of the sign of $s \propto \partial\varepsilon_s/\partial T$ from positive to negative (from negative to positive) [7], as shown in Fig. 3.

As mentioned above, Fröhlich entropy is a macroscopic quantity capable of characterizing the order–disorder state of perovskite ferroelectrics and has the following advantages [8, 9]: (1) Fröhlich entropy can be directly obtained from

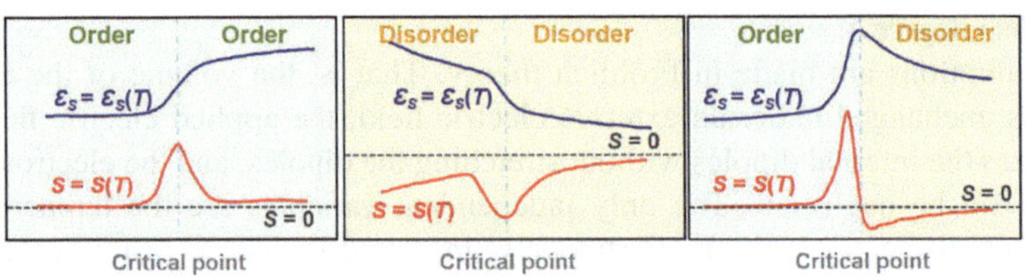

Fig. 3 Schematic temperature dependence of the dielectric constant ε_s (blue curve) and Fröhlich entropy for a unitary field s (red curve) of (**a**) order/order, (**b**) disorder/disorder, and (**c**) order/disorder transitions

dielectric data and can describe the evolution of the structural state with temperature in an almost continuous manner. (2) Because the electric field strength in dielectric tests is extremely low (usually less than 1 V/cm), there is no coercive field effect on the polar nanoregions (PNRs) or local polarization structures in ferroelectrics. This makes it suitable for studying the thermal evolution of a system, especially the phase transition process, in which the system state changes drastically. (3) Fröhlich entropy reflects the effect of the external electric field on entropy. Because the electric field is a physical quantity with a specific direction, applying an electric field in different directions can investigate the specific-direction components of the global entropy and provide information on the directional characteristics of the degree of order in the material. These advantages endow the Fröhlich entropy and $s(T)$ with unique superiority for studying the order–disorder phase transition and evolution characteristics of perovskite ferroelectrics. As a result, they have become an effective supplementary method for fixed-temperature structural characterization techniques, such as X-ray, electron, and neutron diffraction.

3 Unravelling Structural Evolution in Order–Disorder Phase Transition by Fröhlich Entropy

Ferroelectric phase transitions generally fall into two categories: ferroelectric-paraelectric phase transitions and ferroelectric-ferroelectric phase transitions. To some extent, both types of these transitions may manifest features associated with order-disorder components. In this section, each type of these two distinct phase transitions will be introduced individually to offer a comprehensive understanding of the role of Fröhlich entropy within them.

3.1 Ferroelectric–Paraelectric Phase Transition

The phase structure of perovskite ferroelectrics evolves continuously as the temperature changes. The crystals exhibit ferroelectricity only within a certain temperature range. Generally, when the crystal exhibits ferroelectricity, the crystal phase structure state is called the ferroelectric phase, whereas the phase structure state without ferroelectricity is called the paraelectric phase. The transformation from the ferroelectric phase structure to the paraelectric phase structure is known as the ferroelectric–paraelectric phase transition. Consider the $BaTiO_3$ crystal as an example. When the temperature is below 120 °C, the crystal exhibits ferroelectricity. However, when the temperature is above 120 °C, its ferroelectricity disappears. The corresponding lattice structure also changes: below 120 °C, $BaTiO_3$ crystals are in a tetragonal system with a 4 mm point group, while above 120 °C, crystals are in a cubic system with a m3m point group. At a temperature of 120 °C, the appearance or disappearance of the ferroelectricity of the $BaTiO_3$ crystal is accompanied by a change in the crystal structure, which is a ferroelectric–paraelectric phase transition. The temperature at which the ferroelectric–paraelectric phase transition occurs is called the Curie temperature (T_C) or Curie point. Different crystals exhibit distinct phase transition scenarios. The ferroelectric–paraelectric phase transitions of $KNbO_3$ and $PbTiO_3$ occur at 435 and 490 °C, respectively. Some crystals possess only a ferroelectric phase, without a paraelectric phase. This is because before the temperature reaches the phase transition temperature, the crystal has already dissolved or decomposed into other components; thus, the Curie point cannot be observed.

Typically, ferroelectrics do not exhibit one ferroelectric phase. Therefore, apart from the ferroelectric–paraelectric phase transition, ferroelectric–ferroelectric phase transitions can also occur between two different ferroelectric phases. The temperatures corresponding to all phase transitions can be collectively referred to as phase transition temperatures. The typical phase structures of perovskite ferroelectrics include rhombohedral, orthorhombic, tetragonal, and cubic phases. The rhombohedral, orthorhombic, and tetragonal phase structures are ferroelectric, whereas the cubic phase structure is paraelectric. The orientations of spontaneous polarisation vary in different ferroelectric phase structures. In the tetragonal, orthorhombic, and rhombohedral phases, spontaneous polarization mainly results from the displacement of the B-site ions from the centre of the oxygen octahedron along the four-, two-, and threefold axes. In the cubic phase, the B-site ions are located at the centre of the oxygen octahedron and the entire crystal has no spontaneous polarization; thus, it belongs to the paraelectric phase. The ferroelectric phase transition is a general term for the ferroelectric–paraelectric and ferroelectric–ferroelectric phase transitions.

One of the macroscopic characteristics of ferroelectrics is that the electric polarization–electric field (P-$\mathbf{E}$) relationship forms a ferroelectric hysteresis loop in the ferroelectric phase. A microscopic explanation for this is the presence of domain structures in the ferroelectric phase. Ferroelectric domains refer to small regions in a ferroelectric crystal where the electric dipole moments of spontaneous polarization are arranged in the same direction. During the ferroelectric–paraelectric (or paraelectric–ferroelectric) phase transition, the stable ferroelectric domains gradually disappear (or nucleate). Compared with the ferroelectric–ferroelectric phase transition, the change in spontaneous polarization $\mathbf{P}_S$ is more drastic in the ferroelectric–paraelectric phase transition, exhibiting a change from presence to absence and from order to disorder. P_S is a key order parameter for describing ferroelectric phase transitions in perovskite ferroelectrics. According to the definition, the phase in which the order parameter appears is called the ordered phase, and the phase without the order parameter is called the disordered phase. Therefore, the contribution of the order–disorder component is more prominent in the ferroelectric–paraelectric phase transition. The ferroelectric–paraelectric phase transitions of $BaTiO_3$ and $PbTiO_3$ introduced above exhibit order–disorder characteristics. In solid-solution perovskite ferroelectrics, this order–disorder feature is even more pronounced.

A solid solution is a structure in which solute atoms dissolve in a solvent lattice while maintaining the structure of the solvent. In solid-solution perovskite ferroelectrics, the crystal structure retains the basic perovskite ABO_3 configuration, where the A-site and B-site can be occupied by different ions, forming a composite perovskite-type crystal. Owing to the compositional diversity of a solid solution, its physical and chemical properties can be optimized by adjusting its composition. This is one of the most important methods and mainstream approaches for improving the performance of perovskite solid solutions. The chemical disorder of its components also enhances the order–disorder characteristics of the phase transition, making the phase transition process more complex.

In 1983, Burns and Dacol studied the temperature-response characteristics of the refractive index of perovskite solid solutions. They found that, at the Burns temperature (T_B), which is well above T_C, the change in the refractive index deviates from the original theoretical law that the thermally induced refractive index is linearly variable. They suggested that below T_B, local spontaneous polarization (PNRs) occurs within the ferroelectrics [10]. PNRs are nano-sized reconfigurable polar regions, which are mesoscopic regions with inherent electric polarization. They are characterized by small size (usually on the nanometre scale), random orientation, and dynamic variability. PNRs are important correlated disorder functional units in ferroelectrics. They are a significant source of high polarizability and play a crucial role in the performance of ferroelectrics [11]. Owing to the random occupation of the same site by different ions, below T_B, the internal compositional disorder can cause lattice distortion, local strain, or random fields. This is the main reason for the generation of PNRs or local polarization above T_C [12]. During the ferroelectric–paraelectric phase transition or the reverse process, the scale of PNRs, intensity of local dipole moments, and local polarization characteristics all evolve, exhibiting typical features of dipolar liquids, dipolar solids, and even dipolar glasses [9, 11]. In addition to the characteristic temperatures T_B and T_C, there are characteristic temperatures T^* and T_f that satisfy the relationship $T_f < T_C < T^* < T_B$ [13]. These four characteristic temperatures are directly related to the evolution of the polarization state during the ferroelectric–paraelectric phase transition (tetragonal–cubic phase transition). T_f, T^*, and T_B can represent the characteristic temperatures at which PNRs freeze, PNRs strongly correlate, and highly active PNRs appear respectively in the cooling process. This evolution of PNRs and local polarization directly leads to a change in the degree of polarization order. Therefore, the order–disorder characteristics of the ferroelectric–paraelectric phase transition in solid-solution perovskite ferroelectrics are prominent.

As an orientational entropy, Fröhlich entropy can be considered a measure of the orientational degree of local polarization within perovskite ferroelectrics [14]. Based on Fröhlich entropy, the ordered and disordered states of PNRs and ferroelectric domains at various characteristic temperatures during the order–disorder phase transition can be analysed and evaluated. With the help of Fröhlich entropy, researchers have gained a deeper understanding of the ferroelectric–paraelectric phase transition involving the evolution of PNRs. When $T > T_B$, the crystal is in a completely paraelectric phase with no dipole or polarization, resulting in zero

Fröhlich entropy. Local dipoles are generated when the temperature decreases to a range between T_B and T^*. The increasing correlated coupling between dipoles induces dipole clustering, thereby forming PNRs with local polarization intensity. An externally applied electric field can reduce the entropy of the system. Moreover, within this temperature range, both the dipoles and the PNRs are strongly affected by thermal fluctuations. Therefore, $s(T)$ is a nearly constant negative value, indicating the disordered state of the system. As the temperature decreases further to T^*, the thermal perturbation gradually weakens, and small PNRs coalesce into larger polar nanodomains (PNDs) via dipole association. At this point, the orienting effect of the electric field on the PNDs becomes more prominent, promoting a more ordered polarization. The value of $s(T)$ further decreases as the temperature drops. This feature of $s(T)$ reveals the continuous growth of the volume fraction of PNDs with decreasing temperature. Fröhlich entropy decreases rapidly to a minimum and then increases rapidly to a maximum in the vicinity of T_C and $s(T)$ manifests as a curve with a two-peak shape. The sign changes abruptly from negative to positive. The correlated PNDs further forms ferroelectric domains trough a percolative process. The point at which $s(T) = 0$ is the critical point for the order/disorder transition of the crystal, corresponding to the transition temperature. When the temperature drops below T_C, ferroelectric domains (FDs) rapidly form in the crystal and their spontaneous polarizations are arranged in order. An externally applied electric field disrupts this orderly arrangement. Therefore, $s(T)$ is positive and Fröhlich entropy increases. However, when the temperature is further reduced below T_C, owing to the relatively small electric field in the dielectric test, it gradually becomes more difficult to drive increasingly stable spontaneous polarization, which results in a decrease in $s(T)$, as shown in Fig. 3. However, $s(T)$ remains positive. The evolution of the local polarization leads to an order–disorder transition in the system near T_C. When the temperature is slightly higher than T_C, there are liquid-like dipole systems inside the material. The PNRs can transition relatively freely between different equivalent states and are sensitive to the external electric field, with the system in a disordered state. When the temperature is lower than T_C, ferroelectric domain arrangements emerge within the system. The positive value of $s(T)$ indicates that the system is in a long-range ordered state.

Next, considering potassium tantalate niobate ($KTa_{1-x}Nb_xO_3$, KTN) crystal as an example, we will illustrate the important role of Fröhlich entropy in explaining the structural evolution during the ferroelectric–paraelectric phase transition. KTN is a well-known perovskite solid-solution ferroelectric material. It exhibits excellent electro-optical and nonlinear optical properties. Near T_C, the KTN crystal has the largest quadratic electro-optic coefficient reported so far to the best of our knowledge. It enables a non-paraxial propagation of a sub-micrometer-sized beam for over 1000 diffraction lengths and a broadband giant index of refraction across the entire visible spectrum [15, 16]. Owing to its outstanding quadratic electro-optic properties, high-performance optical deflectors, modulators, and other devices have been developed. Numerous studies have shown that KTN's excellent properties originate from the response characteristics of PNRs in crystals [17, 18]. The ferroelectric–paraelectric phase transition of the KTN crystal is similar to that

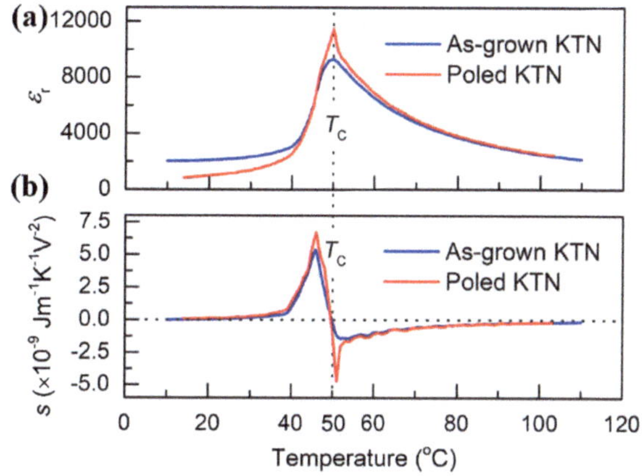

Fig. 4 Macroscopic dielectric responses. (**a**) Relative permittivity ε_r versus temperature of the as-grown and poled samples, respectively. (**b**) Corresponding Fröhlich entropy for a unitary field s. The T_C (50 °C) is marked. Reprinted with permission from [19]. (Courtesy of the original publisher © 2018, Optical Society of America)

of the $BaTiO_3$ crystal, which is a transition from a tetragonal phase structure to a cubic phase structure. The application of Fröhlich entropy clearly reveals the structural evolution during the ferroelectric–paraelectric phase transition (order–disorder transition) of the KTN crystal. The following are two examples.

Poling treatment is one of the commonly used treatment methods for ferroelectric crystals. It is usually used to change the arrangement state of polarization within the ferroelectric-phase crystal. A change in the polarization state leads to changes in the phase transition process. Figure 4 shows a comparison of the dielectric constants and $s(T)$ values of the as-grown and poled KTN samples [19]. In the temperature range close to and above T_C, $s(T)$ undergoes drastic changes. This indicates that the polarization treatment also significantly affects the paraelectric phase structure above T_C. The poling treatment enhances the correlated coupling effect of the local dipoles above T_C, promoting the formation of larger-volume dipole clusters. These larger local polarization units are arranged in a disorderly manner, but with the external electric field, they can be re-oriented in the direction of the electric field, resulting in a significant entropy decrease. When the temperature is below T_C, $s(T)$ is slightly larger than that of the initial sample. This is because, in the ferroelectric phase (tetragonal phase) structure, the poling treatment makes the initial polarization arrangement more orderly. Therefore, the disturbing effect caused by the tested alternating current electric field is more obvious, leading to a greater entropy increase. From the dynamic response perspective, Fröhlich entropy reflects the differences in the polarization states within the crystal after the poling treatment. The dynamic response characteristics of these local polarization units are difficult to determine using conventional static testing methods.

One of the remarkable features of solid-solution crystals is that the composition ratio within the crystal can be artificially adjusted through the design of raw materials and the control of crystal growth. When the growth conditions are strictly controlled, both crystals with a uniform composition and those with a gradient composition can be obtained [20–22]. Fröhlich entropy can assist in understanding the ferroelectric–paraelectric phase transition processes in both uniform and

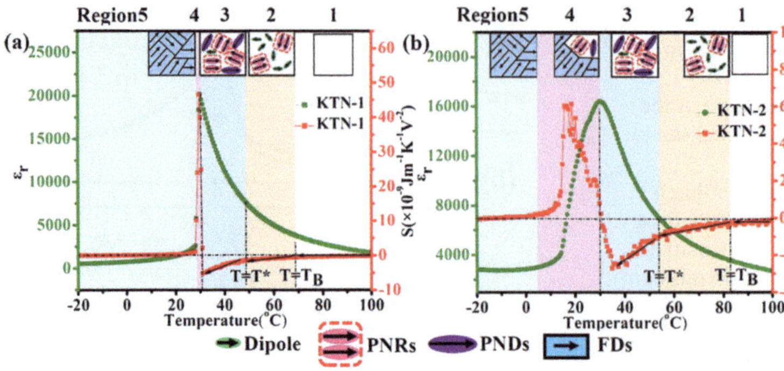

Fig. 5 (**a**) The temperature-dependent properties of ε_r (green) and Fröhlich entropy (red) for the uniform KTN sample at a 1 kHz electric field. (**b**) The temperature-dependent properties of ε_r (green) and Fröhlich entropy (red) for the KTN sample with a large compositional gradient at a 1 kHz electric field. Schematics of structural evolution [dipole, PNRs, PNDs, and ferroelectric domains (FDs)] are inserted corresponding to the temperature ranges. Reprinted with permission from [23]. (Courtesy of the original publisher © 2023, American Chemical Society)

nonuniform samples. Figure 5 compares the $s(T)$ values of KTN crystals with a uniform composition and with a compositional gradient [23]. For a crystal with a uniform composition, the phase transition occurs almost simultaneously throughout the entire crystal. This leads to a sharp and rapid change in the $s(T)$ value near T_C. In the crystal with a compositional gradient, the coexistence of multiple polarization states within the crystal causes a relatively stable trend in the $s(T)$ value in the ferroelectric–paraelectric phase transition temperature range. This situation occurs because the asynchronous phase transitions in different parts of the crystal during the heating (or cooling) process. The compositional gradient delays the progression of the crystal phase transition. The compositional distribution of solid-solution single crystals has a significant impact on both the micro-mesoscopic polarization structure and macroscopic properties of the material. Fröhlich entropy can not only visually determine whether the internal composition of the material is uniform, but also provide ideas for the performance regulation of perovskite ferroelectrics.

3.2 *Ferroelectric–Ferroelectric Phase Transition*

The ferroelectric–ferroelectric phase transitions also contain order–disorder components. Therefore, Fröhlich entropy can be used to study the ferroelectric–ferroelectric phase transition process. Several perovskite ferroelectrics exhibit remarkable anisotropic characteristics. Breaking the lattice symmetry in the ferroelectric phase is the source of anisotropy. During the growth process, the distribution of the temperature field, compositional fluctuations, and the introduction of thermal stress leads to differences in the spontaneous polarization and internal strain in different directions of the crystal [20, 24, 25]. Owing to the intrinsic anisotropy of the ferroelectric phase lattice, this difference is more evident in the ferroelectric phase, which also results in significant directional differences in the ferroelectric–ferroelectric phase transition. Fröhlich entropy is derived from the relationship between the dielectric constant and entropy change. Under an electric field in a certain direction, $s(T)$ can represent the entropy change value of the field-related part in a specific direction [14]. Therefore, Fröhlich entropy is highly directional and can provide directional information about the degree of order of the material, making it suitable for detecting and analysing anisotropic properties in nonuniform systems.

$s(T)$ can be used to measure the Fröhlich entropy component in a specific direction and is used to characterize the arrangement of local dipoles in the solid/liquid (ordered/disordered) states in different directions as the temperature changes. By conducting tests along different directions (a, b, c), a series of dielectric spectra can be obtained, and $s_a(T)$, $s_b(T)$ and $s_c(T)$ can be calculated, which respectively correspond to the component values of $s(T)$ along the a, b and c directions. Parravicini et al. used Fröhlich entropy to study the phase transition differences of KLTN and KNTN samples in the a, b and c directions [8, 9]. Figure 6 shows all the phase transition processes of the KLTN crystal, including the rhombohedral–orthorhombic, orthorhombic–tetragonal, and tetragonal–cubic phase transitions. The first two phase transitions are ferroelectric–ferroelectric phase transition processes. The dielectric responses of the crystal samples in different directions exhibit significant characteristic differences. Moreover, as can be seen from $s(T)$, in all the phase transition processes, an order–disorder phase transition marked by the positive-negative double-peak curve of $s(T)$ occurs in certain directions. This includes the rhombohedral–orthorhombic phase transition in the a direction, the orthorhombic–tetragonal phase transition in the a and b directions, and the tetragonal–cubic phase transition in the a, b, and c directions. Among them, the tetragonal–cubic phase transition, that is, the ferroelectric–paraelectric phase transition, exhibits similar evolution trends along the three directions, which is consistent with the ferroelectric–paraelectric phase transition process described in the previous section. However, for ferroelectric–ferroelectric phase transition processes, the change in Fröhlich entropy shows obvious anisotropy.

In a given direction, each long-range symmetry phase is associated with a specific dipolar solid/liquid state. In the lowest-temperature rhombohedral phase, $s(T)$ is positive in all directions and increases continuously with increasing temperature. This response demonstrates that the rhombohedral phase exhibits stable and fully ordered solid-like state (ordered state) characteristics. In the orthorhombic phase, the crystal exhibits an ordered state with a positive value of $s(T)$ along the b and c directions, whereas along the a direction, $s(T) < 0$, indicating a disordered liquid-like state (orientational liquid feature). As the temperature increases further, the system transforms into the tetragonal phase and is highly disordered in the a and b directions while remaining ordered in the c direction. In the cubic phase, the disordered liquid-like state is observed in all three directions, exhibiting isotropy. The results indicate that the crystal is fully long-range ordered in the

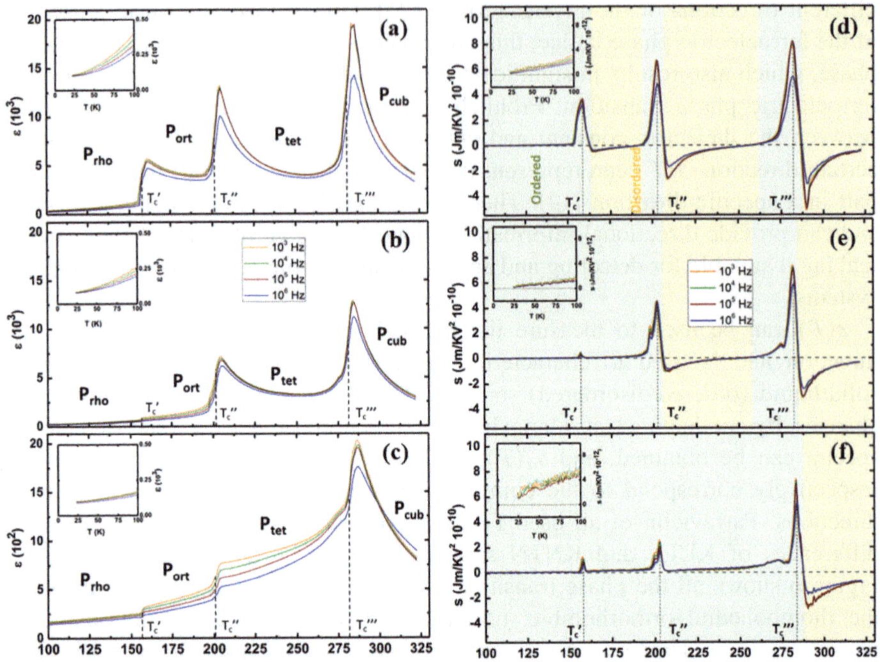

Fig. 6 Dependencies on temperature of the relative permittivity measured at 10^3–10^6 Hz frequencies along *a* (**a**), *b* (**b**), and *c* (**c**) directions for the cooling stage. Fröhlich entropy for the unit of volume and field square s_a, s_b, and s_c for, respectively, *a* (**d**), *b* (**e**), *c* (**f**) axes directions, in the 100–320 K temperature range (cooling stage). The insets show the permittivity behaviour in the temperature range of 0–100 K for each respective direction. Reprinted with permission from [8]. (Courtesy of the original publisher © 2016, American Physical Society)

rhombohedral phase, whereas in the orthorhombic or tetragonal phases, solid-like (ordered) and liquid-like (disordered) dipolar states can coexist in different directions, presenting as a liquid-solid directional composite system, as illustrated in Fig. 7 [9]. This is the reason for the anisotropy observed in the ferroelectric–ferroelectric phase transitions. The conclusions obtained from the Fröhlich entropy analysis regarding the macroscopic order/disorder states of the four-phase structures and three crystallographic directions align well with the eight-sites model developed by Comès et al. The eight-sites model indicates that, in these compounds, an intrinsic disorder is a consequence of the dynamic displacement of the B cation on eight energetically equivalent sites along the diagonals of the cubic cell, whose allowed directions of the displacements are dependent on the given crystalline phase [8, 26]. Specifically, they determined that in the cubic phase, there are three planes of diffusive scattering normal to the three cubic axes (i.e. disorder is manifested in all directions); in the tetragonal phase, there are only two (disorder is manifested in the directions perpendicular to the tetragonal *c* axis); and in the orthorhombic phase, there is only one plane of diffusion. Finally, in the rhombohedral phase,

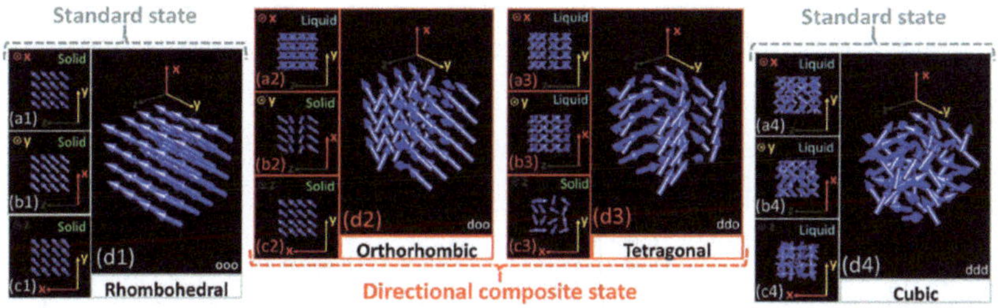

Fig. 7 Schematic representation of the nanometric dipole arrangement in the four nominal phases. On the opposite ends the two standard fully solid or fully liquid states: the rhombohedral phase with three ordered (solid) directions (ordered, ordered, ordered, "ooo"): global view (d1) and projections along the axes x (ordered, solid-like, a1), y (ordered, solid-like, b1), and z (ordered, solid-like, c1); the cubic phase with three disordered (liquid) directions (disordered, disordered, disordered, "ddd"): global view (d4) and projections along the axes x (disordered, liquid-like, a4), y (disordered, liquid-like, b4), and z (disordered, liquid-like, c4). In the middle, the two exotic liquid–solid directional composite states: the orthorhombic phase with two ordered (solid) and one disordered (liquid) directions (disordered, ordered, ordered, "doo"): global view (d2) and projections along the axes x (disordered, liquid-like, a2), y (ordered, solid-like, b2), and z (ordered, solid-like, c2); the tetragonal phase with one ordered (solid) and two disordered (liquid) directions (disordered, disordered, ordered, "ddo"): global view (d3) and projections along the axes x (disordered, liquid-like, a3), y (disordered, liquid-like, b3), and z (ordered, solid-like, c3) [9]. Reprinted with permission from [9]. © 2017, Royal Society of Chemistry

there is no diffusion (only this symmetry shows the behaviour of a true crystal): in this phase, the displacements are correlated and the B cations are displaced in the same direction. Therefore, the rhombohedral phase is found to be the sole ordered and stable phase for the cited perovskites [27–30]. The agreement between the experimental results and eight-sites model strongly suggests that the crystal, as a quasi-single-domain entity, responds to electric fields, and this "self-poling" state may be induced by the strain-triggered flexoelectric effect and other factors [8]. Additionally, the inhomogeneous orientation introduced during growth leads to slight deviations in the experimental results compared to the eight-sites model. For instance, according to the eight-sites model, in the cubic phase, the probability of B-site cation displacements is the same in all three directions; therefore, the amount of disorder along the three crystallographic axes (a, b, c) should be equal. However, the experimental results indicate that for a given temperature T, $s_a(T) < s_b(T) < s_c(T)$, which contradicts the centrosymmetric nature of the cubic phase and suggests the presence of anisotropic characteristics in the cubic phase. In the tetragonal phase, the different levels of disorder in the a and b directions also deviate from the theoretical predictions for the tetragonal phase. These observations provide evidence of strong internal symmetry breaking.

Furthermore, it was observed in the experiments that in any phase and direction where a disordered state is displayed, the $s(T)$ behaviour exhibits similar features. As shown in Fig. 8, the negative $s(T)$ curve shows that upon cooling a continuous decrease to a minimum, then the trend is suddenly inverted through a very sharp increase, and the induced entropy becomes positive during the structural transition

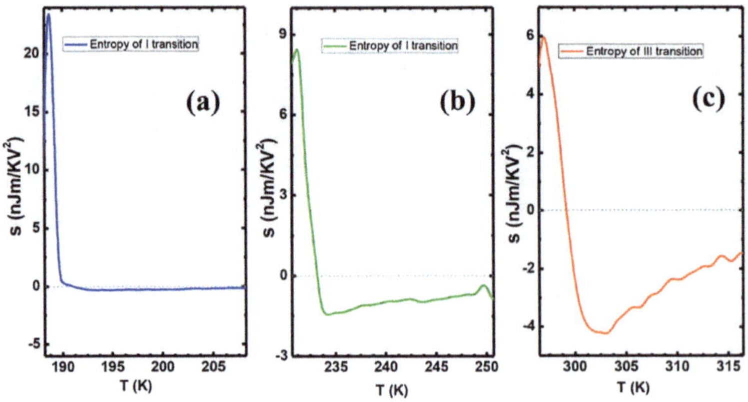

Fig. 8 The a-direction Fröhlich entropy in the temperature right-neighbourhood of the three detected transitions, rhombohedral–orthorhombic (**a**), orthorhombic–tetragonal (**b**) and tetragonal–cubic (**c**), respectively. The reported temperature interval is 20 K in the plots of all the considered transitions [9]. (Courtesy of the original publisher © 2017, Royal Society of Chemistry)

[9]. The rationale behind this trend is akin to previous discussions on PNRs in ferroelectric–paraelectric phase transitions. Research shows that polar nanoregions (PNRs) also exist in the ferroelectric phase structure and exhibit high electric-field response activity near the phase boundary [31, 32]. The $s(T)$ trend indicates that during the ferroelectric–ferroelectric phase transitions, the size of the PNRs in the system also increase with decreasing temperature. Upon reaching a critical temperature, the PNRs lose their freedom of reorientation and freeze to match the symmetry of the low-temperature phase, thereby rendering $s(T)$ positive. Moreover, during the ferroelectric–ferroelectric phase transitions at lower temperatures, the absolute magnitude of the negative $s(T)$ is lower. This behaviour may be attributed to the competition between the displacive and order–disorder characteristics during the phase transition. In phase transitions at lower temperatures, the order–disorder component diminishes, whereas the displacive characteristic becomes increasingly prominent. The evolution of Fröhlich entropy confirms the coexistence of two components (order-disorder and displacive) in the ferroelectric phase transition. The aforementioned studies suggest that Fröhlich entropy is an effective method for investigating macroscopic order/disorder states along different directions within the long-range symmetry phases of perovskite ferroelectric materials. It enables intuitive and nearly continuous observation of order–disorder phase transitions in different directions during temperature variations and provides compelling evidence for symmetry breaking in phase transitions. Therefore, it holds significant value in studies of ferroelectric–ferroelectric phase transitions.

4 Role of Fröhlich Entropy in Revealing the Origin of Properties of Perovskite Ferroelectrics

In ferroelectric materials, the coupling effects among the optical, electrical, mechanical, and thermal properties are ubiquitous, exhibiting rich physical connotations and extensive application value, which makes ferroelectric materials key functional materials in numerous fields, such as information, energy, and mechanics. Their excellent performance is closely related to their internal polarization state. The Fröhlich entropy directly reflects the field-induced dynamic characteristics of PNRs and local dipoles, which aids in revealing the origin of the properties of perovskite ferroelectrics and guides performance optimisation. Below, we elucidate the role of the Fröhlich entropy in understanding and predicting the properties based on several representative effects of perovskite ferroelectrics.

Dielectric Property According to the Fröhlich theory, Fröhlich entropy is derived from the dielectric response; thus, the property most directly reflected by Fröhlich entropy is the dielectric property. In perovskite ferroelectrics, the electric polarizability of local polarization such as PNRs and local dipoles, is a significant source of dielectric performance. The giant dielectric response near the order–disorder transition point is closely related to the dynamic characteristics of local polarization. Rapid cooling was found to enhance the dielectric response of the KTN crystals [33]. Relying on the Fröhlich entropy, we can clearly understand the reasons for the improvement in dielectric performance. After rapid cooling, the Fröhlich entropy above the T_C becomes smaller, indicating that local polarization such as PNRs are intrinsically in a more disordered state. This more disordered state of local polarization possesses a shorter relaxation time τ [34]. Given that the real part of the electric polarizability is $\chi' = \chi_s/[1 + (\omega\tau)^2]$, where χ_s is the static electric polarizability and ω is the angular frequency of the applied electric field, it can be deduced that the disordered state of local polarization leads to higher electric polarizability and a stronger response to external electric fields, which is a crucial factor in enhancing dielectric response. Because the dielectric constant is the cornerstone of ferroelectric material properties, the Fröhlich entropy has high applicability in understanding other properties.

Electro-Optic Effect The electro-optic effect in crystals refers to the phenomenon in which the refractive index of a crystal changes under the application of an external electric field, which enables the modulation of the phase and intensity of optical signals. The contributions of the electrons, ions, and dipole moments within the crystal collectively influence the macroscopic electro-optic properties. Near the order–disorder phase transition point, PNRs are commonly regarded as the essence of disordered polarization units. The PNRs and dipole clusters formed through their associated coupling are crucial structural factors that affect the electro-optic effect around the phase transition. Their electric field response characteristics impart a notable dependence on the electro-optic effect for both the electric field strength and frequency [17, 19]. Considering the quadratic electro-optic effect above

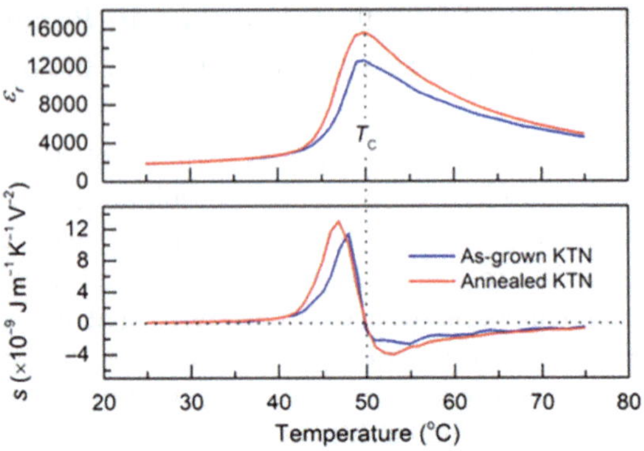

Fig. 9 Temperature dependence of relative permittivity ε_r measured at 1 kHz and Fröhlich entropy. T_C is 50 °C. Reprinted with permission from [12]. (Courtesy of the original publisher © 2019, American Physical Society)

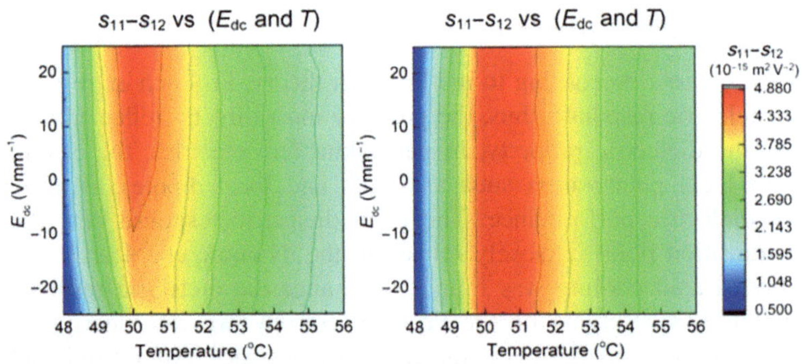

Fig. 10 Electric field and temperature dependences of the quadratic EO coefficient s_{11}-s_{12} of the as-grown sample (left) and the annealed sample (right) under ac + dc electric fields, respectively. The E_{ac} component is $15\,V \cdot m^{-1}$ [12]. Reprinted with permission from [12]. (Courtesy of the original publisher © 2019, American Physical Society)

T_C as an example, the quadratic electro-optic coefficient is proportional to the square of the electric polarizability. Local polarization in a disordered state can lead to higher electric polarizability, thereby enhancing the electro-optic performance. Chang et al. enhanced the electro-optic properties near T_C in KTN crystals by rapid cooling. At a certain cooling rate of $0.45° \cdot s^{-1}$, a remarkable quadratic electro-optic coefficient s_{11}-s_{12} of $6.94 \times 10^{-14}\ m^2 \cdot V^{-2}$ was achieved at T_C [35]. The disordered state of the PNRs, indicated by the smaller Fröhlich entropy resulting from rapid cooling is crucial for this significant property enhancement. Additionally, Tan et al. found that as-grown and annealed KTN crystals exhibited different $s(T)$ values, leading to distinct electro-optic response characteristics, as shown in Figs. 9 and 10 [12]. For as-grown KTN crystals, internal fields often exist in the crystal because of factors such as internal stress, which restrict the disordered dynamics of local polarization. These factors can be eliminated through annealing, which allows the local polarization activity to recover, resulting in a more disordered local polarization corresponding to a smaller $s(T)$ above T_C. The more disordered polarization state of PNRs above T_C after annealing led to a nearly twofold increase in the s_{11}-s_{12}, from 2.57×10^{-15} $m^2 \cdot V^{-2}$ to 4.85×10^{-15} $m^2 \cdot V^{-2}$ [12]. Furthermore, owing to the disappearance of the internal field, the original asymmetric distribution of the electro-optic coefficients under a biased electric field became uniform. Within the range of the applied bias electric field, the influence of the bias electric field on the electro-optic coefficient vanished. Therefore, the utilization of Fröhlich entropy aids in understanding the enhancement mechanism of electro-optic properties near the order–disorder phase transition in perovskite ferroelectrics.

Electrocaloric Effect The electrocaloric effect refers to the adiabatic temperature change or isothermal entropy change resulting from alterations in the polarization state induced by changes in the external electric field in polar materials, which can be utilized for solid-state refrigeration [36–38]. The isothermal entropy change associated with the electrocaloric effect is expressed as:

$$\Delta S = \int_{E_1}^{E_2} \left(\frac{\partial \mathbf{D}}{\partial T}\right)_{\mathbf{E}} \cdot d\mathbf{E} \tag{5}$$

where T is the ambient temperature, $\mathbf{D}$ is the electric displacement vector, and $\mathbf{E}$ is the electric-field strength. Although this isothermal entropy change also involves the influence of an electric field, it is entirely distinct from Fröhlich entropy. In the electrocaloric effect, the applied electric field is often relatively large, exceeding the coercive field, which introduces entropy changes due to the reversal, reorientation, and contraction/expansion of the ferroelectric domains or nanodomains. In contrast, Fröhlich entropy reflects the field-induced characteristics of disordered local polarization without coercive field effects. Nevertheless, Fröhlich entropy retains significant value in understanding the electrocaloric effect. A drastic change in the polarization intensity near the ferroelectric phase transition often leads to exceptional electrocaloric performance. Huang et al. employed Fröhlich entropy to investigate the mechanism by which anisotropic polarization responses near the ferroelectric–paraelectric phase transition enhance the electrocaloric effect in perovskite ferroelectrics. By characterizing the Fröhlich entropy of the KTN crystals along the $[111]_C$, $[011]_C$, and $[001]_C$ crystallographic directions, it was found that the polarization state along the $[111]_C$ direction exhibits higher electric field response activity, as illustrated in Figs. 11 and 12. Moreover, when a large electric field is applied along the $[111]_C$ direction and subsequently removed, the microdomain structure with tetragonal symmetry can avoid being poled and recover a considerably disordered arrangement, which facilitates a larger isothermal entropy change and achieves a better refrigeration effect. At an electric field of 15 $kV \cdot cm^{-1}$, the adiabatic temperature change ΔT of KTN along the $[111]_C$ direction at T_C increases by nearly 120% compared to ΔT along the $[001]_C$ direction, from 0.15 to 0.33 K [36].

Electric-Field-Induced Strain Electric-field-induced strain refers to the phenomenon in which a dielectric undergoes elastic deformation under an electric field. In general, this effect is relatively weak. However, for ferroelectrics, the electric-

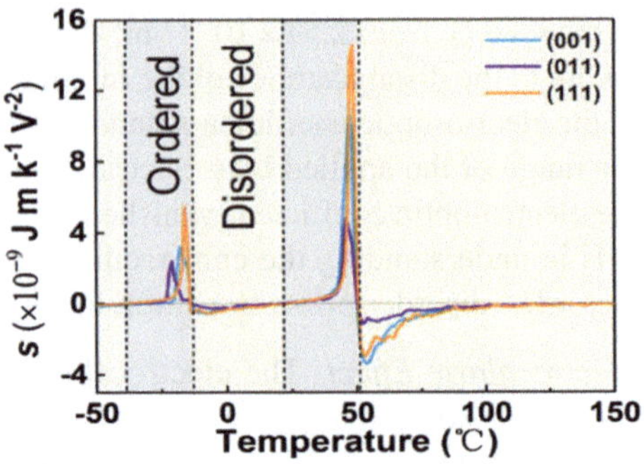

Fig. 11 Changes of $s(T)$ along $[001]_C$, $[011]_C$, and $[111]_C$ directions of KTN. The temperature range is from −50 to 150 °C [36]. Reprinted with permission from [36]. (Courtesy of the original publisher © 2022, American Chemical Society)

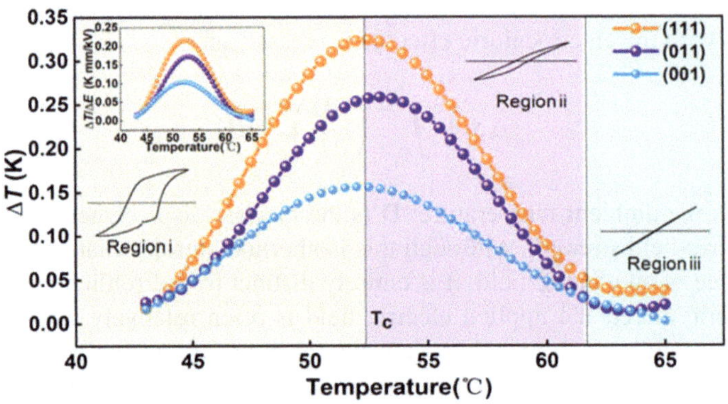

Fig. 12 Adiabatic temperature change (ΔT) of anisotropic samples along different directions. The inset shows the electrocaloric strength ($\Delta T/\Delta \mathbf{E}$) of the three samples near the ferroelectric-paraelectric phase transition. The **P-E** loops display the different states in Regions i–iii [36]. Reprinted with permission from [36]. (Courtesy of the original publisher © 2022, American Chemical Society)

field-induced strain can reach relatively high values (10^{-2}–10^{-3}). The reversal of ferroelectric domain structures inside ferroelectrics under the electric field is a key source of the excellent performance [39]. In ferroelectrics, the generation of internal fields is caused by various factors such as defects, internal stresses, and composition gradients. The restraining effect of the internal field on the ferroelectric domains suppresses the manifestation of electric-field-induced strain. The Fröhlich entropy can reflect the response characteristics of the local polarization under a weak electric field, thus indicating the strength of the internal field. Moreover, its directionality can be used to clarify the distribution characteristics of internal fields in ferroelectrics. For example, in KTN crystals, experiments show a clear correspondence between the Fröhlich entropy and the composition gradient, as shown in Fig. 13. In the direction with a larger composition gradient in the crystal, the internal dipole moments exhibit a more ordered state; that is, the value of $|s(T)|$ near the phase transition point becomes significantly smaller. This proves the existence of a strong internal field acting along the direction of the large composition

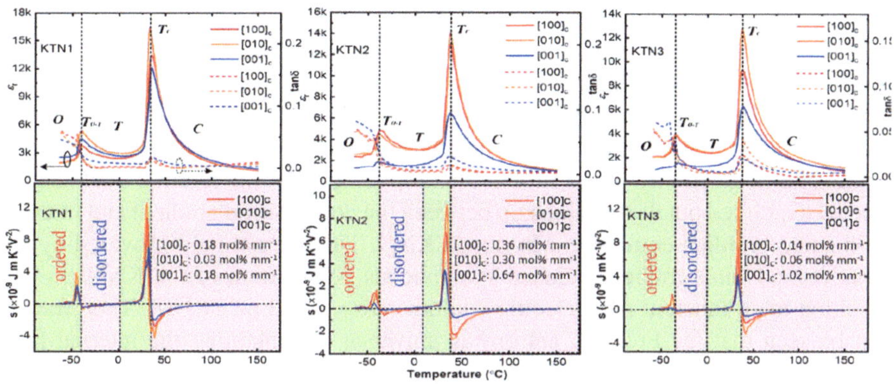

Fig. 13 Temperature dependence of dielectric permittivity (full line) and dielectric losses (dashed line) along three crystallographic directions measured at 1 kHz under the cooling stage for KTN1, KTN2, and KTN3 [24]. Reprinted with permission from [20]. (Courtesy of the original publisher © 2020, Elsevier)

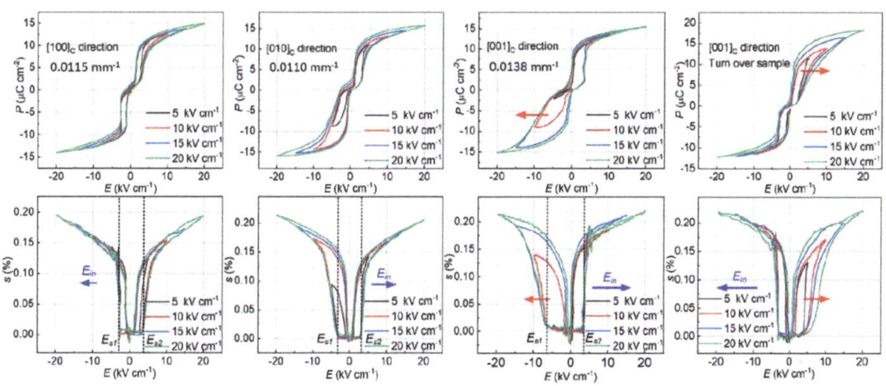

Fig. 14 The bipolar **P**-**E** loops and strain curves of samples (cut from KTN4) under different electric fields along $[100]_C$, $[010]_C$, and $[001]_C$, respectively [20]. The component gradients along different directions of these samples are inserted in the figures. Reprinted with permission from [20]. (Courtesy of the original publisher © 2020, Elsevier)

gradient. During the ferroelectric phase transition, a strong internal field can restrict the originally ordered polarization orientation when the temperature is lower than the phase transition temperature. This makes it difficult for the external electric field in the dielectric tests to drive local polarization. Consequently, the value of $|s(T)|$ is smaller. When the temperature is higher than the phase transition temperature, the internal field confines the local polarization, causing the polarization to tend towards a more ordered state. This leads to a reduction in disordered dipoles, which cannot generate a sufficiently strong response. Consequently, the value of $|s(T)|$ decreases. As shown in Fig. 14, $\mathbf{P} - \mathbf{E}$ loops and strain curves confirm the conclusions drawn from the Fröhlich entropy.

In the direction with a smaller composition gradient, $\mathbf{P}-\mathbf{E}$ loops exhibit good symmetry. However, in the direction with a large composition gradient, the $\mathbf{P}-\mathbf{E}$ loops exhibit more distinct asymmetry. A similar phenomenon is observed in the strain curve, that is, the $S-\mathbf{E}$ curve [20]. This asymmetry is evidence of the existence of an internal field in the direction of the composition gradient, which demonstrates the ability of the Fröhlich entropy to predict the internal field. Zhang et al. also pointed out the relationship between the composition gradient and internal field using Fröhlich entropy in Mn-doped KNN. The internal field provided by the large composition gradient induced spontaneous polarization of the KNN crystal, resulting in an inherent piezoelectric coefficient d_{33} of 83 pC/N [40]. Therefore, it can be seen that the Fröhlich entropy is universal in explaining the internal-field factors that influence electric-field-induced strain and the piezoelectric coefficient.

In summary, Fröhlich entropy can not only reveal the evolution process of the order–disorder phase transition in perovskite ferroelectrics but also provide a powerful tool for understanding and predicting the properties of perovskite ferroelectrics from the perspective of order–disorder transformations. It supplements the structural and property characterization techniques and presents a more comprehensive physical picture. In principle, all electric-dipole-related evolution processes and properties in perovskite ferroelectrics involve the transformation between ordered and disordered states, which can potentially be studied using the non-destructive and easily characterizable Fröhlich entropy method.

References

1. L. Landau, E. Lifshitz, *Statistical Physics*, 3rd edn. (Pergamon, Oxford, 1980)
2. T. Mitsui, I. Tatsuzaki, E. Nakamura, *An Introduction to the Physics of Ferroelectrics* (Gordon and Breach, New York, 1976)
3. K. Itoh, L. Zeng, E. Nakamura, N. Mishima, Ferroelectrics **63**(1), 29 (1985)
4. R. Nelmes, R. Piltz, W. Kuhs, Z. Tun, R. Restori, Ferroelectrics **108**(1), 165 (1990)
5. Z. Lei, X. Liu, Y. Wu, et al., Nature **563**(7732), 546 (2018)
6. C. Oses, C. Toher, S. Curtarolo, Nat. Rev. Mater. **5**(4), 295 (2020)
7. H. Fröhlich, *Theory of Dielectrics: Dielectric Constant and Dielectric loss*, 2nd edn. (Claredon Press, Oxford, 1958)
8. J. Parravicini, E. DelRe, A. Agranat, G. Parravicini, Phys. Rev. B **93**(9), 094203 (2016)
9. J. Parravicini, E. DelRe, A. Agranat, G. Parravicini, Nanoscale **9**(27), 9572 (2017)
10. G. Burns, F. Dacol, Phys. Rev. B **28**(5), 2527 (1983)
11. J. Parravicini, E. DelRe, S. Perego, et al., Phys. Rev. B **106**(6), 064107 (2022)
12. P. Tan, H. Tian, F. Huang, et al., Phys. Rev. Appl. **11**(2), 024037 (2019)
13. J. Toulouse, Ferroelectrics **369**(1), 203 (2008)
14. J. Parravicini, G. Parravicini, Results Phys. **28**, 104571 (2021)
15. E. DelRe, F. Di Mei, J. Parravicini, et al., Nat. Photon. **9**(4), 228 (2015)
16. F. Di Mei, L. Falsi, M. Flammini, et al., Nat. Photon. **12**(12), 734 (2018)
17. P. Tan, H. Tian, C. Mao, et al., Appl. Phys. Lett. **111**(1) (2017)
18. J. Zhang, X. Du, X. Wang, et al., Appl. Phys. A **127**(5), 386 (2021)
19. P. Tan, H. Tian, Y. Wang, et al., Opt. Lett. **43**(20), 5009 (2018)
20. F. Huang, C. Hu, Z. Zhou, et al., Acta Mater. **200**, 24 (2020)
21. C. Hu, X. Meng, M. Zhang, et al., Sci. Adv. **6**(13), eaay5979 (2020)

22. B. Liu, H. Zhang, Y. Zhang, et al., Acta Phys. Pol. A **135**(3) (2019)
23. P. Chen, J. Zhao, X. Du, et al., Cryst. Growth Des. **23**(7), 5236 (2023)
24. F. Huang, C. Hu, H. Tian, et al., Cryst. Eng. Comm. **21**(45), 7002 (2019)
25. F. Huang, C. Hu, H. Tian, et al., Cryst. Growth Des. **19**(9), 5362 (2019)
26. J. Parravicini, G. Parravicini, Preprint. arXiv:2104.07564 (2021)
27. R. Comes, M. Lambert, A. Guinier, Solid State Commun. **6**(10), 715 (1968)
28. R. Comes, M. Lambert, A. Guinier, Found. Crystallogr. **26**(2), 244 (1970)
29. K. Tsuda, R. Sano, M. Tanaka, Phys. Rev. B Condens. Matter Mater. Phys. **86**(21), 214106 (2012)
30. K. Tsuda, R. Sano, M. Tanaka, Appl. Phys. Lett. **102**(5), 051913 (2013)
31. F. Li, S. Zhang, T. Yang, et al., Nat. Commun. **7**(1), 13807 (2016)
32. F. Li, D. Lin, Z. Chen, et al., Nat. Mater. **17**(4), 349 (2018)
33. J. Parravicini, C. Conti, A. Agranat, E. DelRe, Opt. Lett. **37**(12), 2355 (2012)
34. A. Bokov, Z. Ye, J. Adv. Dielectr. **2**(02), 1241010 (2012)
35. Y. Chang, C. Wang, S. Yin, et al., Opt. Lett. **38**(22), 4574 (2013)
36. X. Huang, P. Tan, Y. Wang, et al., Cryst. Growth Des. **22**(5), 3485 (2022)
37. F. Huang, H. Tian, X. Meng, et al., Phys. Status Solidi (RRL)–Rapid Res. Lett. **13**(2), 1800515 (2019)
38. Z. Kutnjak, B. Rožič, R. Pirc, Wiley Encyclopedia of Electrical and Electronics Engineering (Wiley, Hoboken, 1999), pp. 1–19
39. X. Ren, Nat. Mater. **3**(2), 91 (2004)
40. Y. Zhang, C. Hu, Y. Wang, et al., J. Mater. Chem. C **11**(14), 4837 (2023)

Index

J. Parravicini (ed.), *Fröhlich Entropy Estimation and Dielectric Calorimetry*, Advances in Dielectrics, https://doi.org/10.1007/978-3-032-21454-6

MIX
Papier aus verantwortungsvollen Quellen
Paper from responsible sources
FSC® C105338

If you have any concerns about our products,
you can contact us on
ProductSafety@springernature.com

In case Publisher is established outside the EU,
the EU authorized representative is:
Springer Nature Customer Service Center GmbH
Europaplatz 3, 69115 Heidelberg, Germany

Printed by Libri Plureos GmbH
in Hamburg, Germany